603835

D1760486

Geotechnical Design to Eurocode 7

Springer
London
Berlin
Heidelberg
New York
Barcelona
Hong Kong
Milan
Paris
Santa Clara
Singapore
Tokyo

Trevor L.L. Orr and Eric R. Farrell

Geotechnical Design to Eurocode 7

Springer

Trevor L.L. Orr, PhD
Department of Civil, Structural and Environmental Engineering, University of
Dublin, Trinity College, Dublin 2, Ireland

Eric R. Farrell, PhD
Department of Civil, Structural and Environmental Engineering, University of
Dublin, Trinity College, Dublin 2, Ireland

ISBN-13: 978-1-4471-1206-8   e-ISBN-13:978-1-4471-0803-0

DOI: 10.1007/978-1-4471-0803-0

Springer-Verlag London Berlin Heidelberg
British Library Cataloguing in Publication Data
Orr, T. L. L. (Trevor L. L.)
    Geotechnical design to eurocode 7
    1.Rock mechanics – Standards – Europe 2.Foundations –
    Standards – Europe 3.Soil mechanics – Standards – Europe
    I.Title II.Farrell, Eric R.
    624.1'5'02184
ISBN-13: 978-1-4471-1206-8   e-ISBN-13:978-1-4471-0803-0

Library of Congress Cataloging-in-Publication Data
Orr, Trevor L. L., 1950-
    Geotechnical design to Eurocode 7 / Trevor L.L. Orr and Eric R.
    Farrell.
        p.   cm.
    Includes bibliographical references and index.
ISBN-13:978-1-85233-038-5   e-ISBN-13:978-1-4471-0803-0
    1.Engineering geology--Standards--Europe. 2. Structural design-
    -Standards--Europe. 3. Standards, Engineering--Europe.
    I. Farrell, Eric R., 1945- . II. Title.
    TA705.4.E85O77 1999   98-51614
    624'.02'184--dc21   CIP

Typesetting: Camera ready by authors

69/3830-543210 Printed on acid-free paper

Preface

The purpose of this book is to explain the philosophy set out in Eurocode 7, the new European code of practice for geotechnical design, and, by means of series of typical examples, to show how this philosophy is used in practice. This book is aimed at:

- practising engineers, to assist them to carry out geotechnical designs to Eurocode 7 using the limit state design method and partial factors;
- lecturers and students on courses where design to Eurocode 7 is being taught.

It is envisaged that practising engineers, using this book to assist them carry out geotechnical designs to Eurocode 7, will have access to the prestandard version of Eurocode 7, ENV 1997-1, so the authors have concentrated on the main principles and have not provided a commentary on all the clauses. However sufficient detail has been included in the book to enable it to be used on its own by those learning the design principles who may not have access to Eurocode 7. For example, the values of the partial factors and the principal equations given in Eurocode 7 have been included and these are used in the design examples in this book. To assist the reader, the numbering, layout and titles of the chapters closely follow those presented in Eurocode 7.

The publication of Eurocode 7 marks a very significant development in civil engineering as it introduces, for the first time in most western European countries, the use of the limit state design method with partial factors for geotechnical design. The vision of producing a Eurocode for geotechnical design should be credited to Professor Kevin Nash, a graduate of Trinity College, Dublin and the authors' Alma Mater. As Secretary General of the International Society for Soil Mechanics and Foundation Engineering (ISSMFE), he saw that harmonised Eurocodes, based on the limit state design method, were being prepared by the European Commission for structural design using materials such as concrete and steel, but not for soil. In 1981 he requested Dr. Niels Krebs Ovesen of Denmark to form a committee to draft a limit state code for geotechnical design. At that time Denmark was the only EC country with a limit state code for geotechnical design. Dr. Krebs Ovesen formed a committee consisting of representatives from the member countries of the European Communities and this committee prepared a model limit state code for geotechnical design, the precursor of Eurocode 7, which was published in 1987.

After publication of the model code, work proceeded, again under the leadership of Dr. Krebs Ovesen, on the preparation of the Eurocode for geotechnical design. This was published in 1994 as the prestandard version of Eurocode 7: Part 1, ENV 1997-1, which forms the basis for this book. During the preparation of ENV 1997-1, the work on the Eurocodes was transferred from the European Commission to CEN, the European Committee for Strandardization. The prestandard version of

Eurocode 7 has now been used for a trial period. Based on the comments received during this trial period, the prestandard is being converted into a full European standard, EN 1997-1, which, it is anticipated, will be published in 2003. Dr. Roger Frank, from France, has taken over the direction of this work from Dr. Krebs Ovesen. Professor Ulrich Smoltczyk from Germany is convenor of the sub-committees redrafting the prestandard and converting it into a full standard.

The concept of changing from the traditional design method, with overall factors of safety, to designs based on the limit state design method and partial factors has presented difficulties for many engineers. There has been much scepticism among certain parts of the geotechnical community about Eurocode 7 and much debate about the suitability of the limit state method, characteristic values and partial factors for geotechnical designs. However, the rational basis of the limit state design method and the advantages of its use in geotechnical design have become more evident as the concepts in Eurocode 7 have matured and been tested. In writing this book, the authors hope to demonstrate how, through being both rational and methodical, the design philosophy presented in Eurocode 7 provides a logical structure that helps clarify and improve the geotechnical design process. It also serves to dispel the image of geotechnical design as a mysterious art and enables it to become a rigorous part of engineering design.

It is clear that the design philosophy presented in Eurocode 7 is having a major influence on geotechnical design, not only in Europe but worldwide. Geotechnical engineers from countries such as Hungary, Israel, Japan, New Zealand and South Africa, to mention but a few, have either modelled their new geotechnical codes on Eurocode 7 or else have taken a very close interest in the development of Eurocode 7. It is likely that this interest will continue and therefore the authors have written this book for non-European as well as European readers.

Acknowledgements

The authors wish to record their thanks to the many colleagues with whom they have worked for many years on the various Eurocode committees and whose support to the authors and different contributions to the development of Eurocode 7 have made this book possible. In particular the authors would like to thank Christophe Bauduin, Ulf Bergdahl, Andrew Bond, Leendert Buth, Richard Driscoll, Roger Frank, Wim Heijnen, Niels Krebs Ovesen, Manuel Matos Fernandes, Willi Sadgorski, Hans Schneider, Bernd Schuppener, Brian Simpson and Ulrich Smoltczyk. The authors also wish to thank Sean Davitt, Martin Grace, Barry Lehane and Tim Paul for their comments on the drafts, Eoin Dunne and Dermot O'Dwyer for their assistance with the drawings, and Simon Perry, Head of Department.

The authors acknowledge with gratitude the support and endurance of their wives, children, extended families and colleagues during the writing of this book.

Trevor Orr
Eric Farrell

Trinity College, Dublin
March, 1999

Contents

Chapter 1

Introduction

1.1 Basis of this Book

This book is based on the document ENV 1997-1 (1994), which is Eurocode 7 - Part 1: Geotechnical Design - General Rules published in 1994 by CEN, the Comité Européen de Normalisation. Hereafter, throughout this book, this document is referred to as EC7. References in this book to the clauses of EC7 are written in bold text and preceded by the letter C, e.g. **C1.2**, while references to other sections of this book are written in normal text and preceded by the symbol §, e.g. §1.2.

Following its publication in 1994 as an ENV, EC7 has been through a trial period, during which comments have been received by CEN from the different European national standards organisations. These comments have been discussed by the CEN committees responsible for converting EC7 into a full European Standard (EN) which is due to be published in 2003. The authors of this book have been closely involved in these discussions and in the developments that have occurred during the trial period. Some of the changes likely to be introduced in the EN version of EC7 as a result of these developments include a clearer definition of the characteristic value, more design cases and a new section on hydraulic failure. The authors have attempted to anticipate these changes in this book.

1.2 The Eurocodes and Eurocode 7

The Eurocodes were conceived by the European Commission in the 1970s as a group of harmonised European standards for the structural and geotechnical design of buildings and civil engineering works. They are concerned with the requirements for the strength, stability, serviceability and durability of structures.

One of the objectives of the Eurocodes is the removal of the barriers to trade in the construction industry in Europe which occur as a result of the existence of different national codes of practice for structural design in the various countries. Although the Eurocodes are European in origin, the concept of a harmonised group

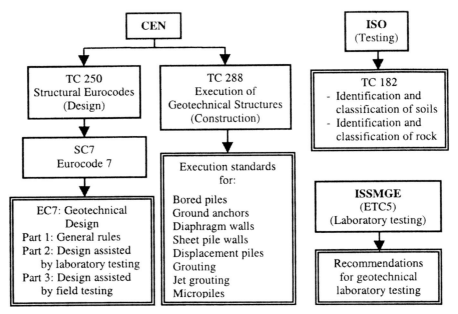

Figure 1.1: *Sources of international standards for geotechnical design, construction and testing*

of codes of practice is relevant worldwide and the Eurocodes could be used as the basis for structural designs throughout the world.

Work on the Eurocodes started in 1976, and on Eurocode 7 in 1981. In 1990 the work on the Eurocodes was transferred from the European Commission to CEN under the direction of the technical committee TC250 Structural Eurocodes, with EC7 being prepared by Sub-Committee 7 (SC7), as shown in Figure 1.1. At present the following nine Eurocodes are being prepared by TC250 covering the basis of design, the actions on structures and design using different materials:

EN 1991 Eurocode 1 (EC1) Basis of design and actions on structures
EN 1992 Eurocode 2 (EC2) Design of concrete structures
EN 1993 Eurocode 3 (EC3) Design of steel structures
EN 1994 Eurocode 4 (EC4) Design of composite and steel and concrete structures
EN 1995 Eurocode 5 (EC5) Design of timber structures
EN 1996 Eurocode 6 (EC6) Design of masonry structures
EN 1997 Eurocode 7 (EC7) Geotechnical design
EN 1998 Eurocode 8 (EC8) Design of structures for earthquake resistance
EN 1999 Eurocode 9 (EC9) Design of aluminium alloy structures

Each Eurocode consists of a number of parts. These Eurocodes, and their different parts, are published by CEN, Brussels. The Secretariat for all the work on EC7 is the Nederlands Normalisatie-instituut (NNI), Delft, Netherlands.

The other Eurocodes that are most relevant to EC7 are:

EC1, which has been split into two separate Eurocodes: *EC0 - Basis of design*, and EC1 - *Actions on structures*. *EC0* sets out the limit state design method with partial factors upon which all the Eurocodes are based.

EC2 - Part 3: Design of concrete structures - Concrete foundations

EC3 - Part 5: Design of steel structures - Piling

EC8 - Part 3: Design of structures for earthquake resistance - Foundations, retaining structures and geotechnical aspects. It should be noted that EC7 does not cover the special requirements of seismic design and hence reference to EC8 Part 3 is necessary for situations involving of earthquake loading.

The Eurocodes are codes of practice for design and so do not provide standards for testing. Geotechnical design, however, is different from most structural designs as the ground material involved is natural, not manufactured like concrete or steel, and so the need to determine the properties of the ground is part of the design process. For this reason Parts 2 and 3 of EC7 have been prepared which, while not being testing standards, cover the requirements for geotechnical design assisted by laboratory and field testing by providing standards for:

- planning geotechnical investigations;
- specifying the requirements for the most commonly used laboratory and field tests;
- interpreting test results and obtaining geotechnical data and the values of geotechnical parameters.

The versions of these documents referred to in this book are, for Part 2, *ENV 1997-2 (1999): Geotechnical design assisted by laboratory testing* and, for Part 3, *ENV 1997-3 (1999): Geotechnical design assisted by field testing*.

1.3   Other Relevant Standards

1.3.1   Standards for Testing

The preparation of standards for testing, whether for soil, concrete or any other material, is the responsibility of other CEN technical committees, not TC250, which, as noted in §1.3, is only concerned with standards for structural design. At present there are no CEN committees preparing standards for soil tests. There is an ISO (International Standards Organisation) committee, TC 182 - Geotechnics in Civil Engineering, which is preparing standards ISO 14688 and 14689 for *Identification and classification of soils* and *Identification and classification of rock*. To avoid duplication, CEN does not normally set up a committee when an ISO committee is already preparing a standard in a particular area. Hence no CEN committee has been established to prepare standards for geotechnical tests.

A European Technical Committee, ETC5, set up by the International Society for Soil Mechanics and Geotechnical Engineering, has prepared recommendations for the most common geotechnical laboratory tests (ISSMGE, 1999). In the absence of European or international standards for geotechnical tests, Parts 2 and 3 of EC7 refer to different national standards for the various laboratory and field tests. These international testing standards and relevant committees are shown in Figure 1.1.

1.3.2 Standards for Construction

A CEN technical committee, TC288 - Execution of Special Geotechnical Works, has been established to prepare standards for the carrying out of geotechnical construction work. As indicated in Figure 1.1, this committee has prepared the following documents, some of which, in spring 1999, are European Standards (EN) while others are prestandards (prENV):

EN 1536   *Execution of special geotechnical works – Bored piles*
EN 1537   *Execution of special geotechnical works – Ground anchors*
EN 1538   *Execution of special geotechnical works – Diaphragm walls*
EN 12063   *Execution of special geotechnical works – Sheet pile walls*
EN 12699   *Execution of special geotechnical works – Displacement piles*
EN 12715   *Execution of special geotechnical works – Grouting*
EN 12716   *Execution of special geotechnical works – Jet grouting*
prEN 288008 Execution of special geotechnical works – Micropiles.

Additional execution standards, for example for reinforced soil and for soil stabilisation, are planned by TC288. Another CEN execution standard, by TC229 – Precast Concrete Products, is *prEN 12794: Precast concrete foundation piles*. As in the case of the Eurocodes, these execution standards are published by CEN, Brussels. The Secretariat for TC288 is the French National Standards Organisation, AFNOR, Paris.

As these execution standards are standards for the carrying out of construction work, they refer to EC7 for design aspects. However, some of them, for example the standards for anchors and diaphragm walls, include design requirements that should be in EC7, but were omitted when the ENV version was written. These requirements will be probably be covered by EC7 when it is revised as an EN. Also, as EC7 was written before the execution standards, it does not refer to them.

1.4 Principles, Application Rules and Assumptions

1.4.1 Principles and Application Rules

A distinction is made in EC7 between clauses that are Principles and those that are Application Rules [**C1.3(1)P**]. Principles, which are indicated by the letter P after the clause number, are general statements, definitions and requirements for which there is no alternative; i.e. they must be satisfied in designs to EC7. Principles are written with the verb "**shall**". Application Rules are examples of generally recognised rules and analytical models that satisfy the Principles. Alternative rules and analytical models may be used provided it is shown that these satisfy the relevant Principles. Application Rules are written with the verb "**should**".

1.4.2 Assumptions

EC7 lists in **C1.4(1)P** a number of assumptions, which are effectively requirements, made concerning designs carried out to EC7, the most important of which are:
- Structures are designed by appropriately qualified and experienced personnel. As no definitions of appropriately qualified person are provided in EC7, some examples are given in §2.2.3.

- Structures are constructed according to the design specifications and relevant standards by appropriately qualified and experienced personnel.
- Adequate continuity and communication exists between the personnel involved in data collection, design and construction. The implications and significance of this assumption are discussed in §4.1.
- There is adequate supervision on site and the completed structure is adequately maintained. These assumptions are discussed in §4.2 and §4.4.

1.5   Terminology

Some of the terminology used in EC7 may appear unfamiliar and strange. Most of the terms are taken from EC1, and so have also been used in all the other Eurocodes, but some terms are specific to EC7 and geotechnical designs. The following definitions are provided to explain this terminology:

Action: Force or imposed displacement that is a known quantity in a calculation model [**C2.4.2(1)P**].

Comparable experience: This is defined in **C1.5.2(1)P** as documented or other clearly established information related to the ground being considered in design, involving the same types of soil and rock and for which similar geotechnical behaviour is expected, and involving similar structures. The importance of information gained locally is emphasised. Comparable experience has been used in place of the more common term 'engineering judgement', which was considered too vague and hence inadequate.

Design: The process of decisions by which the geometry and nature of man-made structures and construction processes are determined. Geotechnical designs to EC7 include all such decisions up to completion of construction and include any related checking and testing, and the specification of any monitoring and maintenance requirements for the completed structure.

Earth pressure: This is the pressure from soil, fill and soft and weathered rocks, and includes the pressure from groundwater.

Execution: The carrying out of construction work on site.

Ground: Soil, rock and fill existing in place before construction starts [**1.5.2(1)P**].

Overall stability: A failure mechanism that involves the whole mass of ground containing the structure.

Structure: Organised combination of connected parts designed to provide some measure of rigidity. This includes fill placed during construction [**C1.5.2(1)P**].

Other terms are defined in the book where they are first used. Many of the clauses in EC7 state that certain factors or requirements shall be either: considered, taken into account, assessed, or evaluated. What exactly is meant by these verbs is not defined in EC7, so the authors suggest the following definitions:

To consider is to think carefully and rationally about the effect of a factor on the design and to decide, on the basis of the available information, what effect it is likely to have. Considering a factor may or may not involve calculations. To ensure that all the relevant factors are considered, the authors propose that lists are drawn up and the items are ticked off as they are considered. Many of the tables in this book, for example Table 2.4, are lists of the items to be considered in designs to EC7.

To take into account is to include the influence or effect of a factor in the design process. In EC7, this verb generally has a stronger meaning than **to consider**.

To assess is to use a process involving some combination of calculations, measurements and comparable experience, to obtain the value of a parameter or check if certain criteria are satisfied.

To evaluate is to obtain the value of a parameter by measurement or calculation, for example the soil strength, a load or the deformation of a structure.

References

ISSMGE (1999) *Recommendations of the ISSMGE for geotechnical laboratory testing*, DIN, Beuth Verlag, Berlin.

Chapter 2

Basis of Geotechnical Design

2.1 Introduction

The basis of geotechnical designs to EC7 is the limit state design philosophy, the main features of which are set out in Section 2 of EC7. These features include the fundamental requirements to be satisfied, the factors to be considered in design, the selection of loads and material parameters, the partial factors on loads and material parameters, and the calculations for checking the ultimate and serviceability limit states. The requirements for presenting the results of a geotechnical design in a Geotechnical Design Report are also given in Section 2.

2.2 Design Principles

2.2.1 Fundamental Requirements and the Design Process

The fundamental requirements to be satisfied by all structures designed to the Eurocodes, including structures designed to EC7, are given in **C2.1(1)P** of EC1 - Part 1 (ENV 1991-1, 1994). These requirements are that a structure shall be designed and executed in such a way that, during its intended life with appropriate degrees of reliability and in an economic way, it will:

- remain fit for the use for which it is required; and
- sustain all actions and influences likely to occur during execution and use.

According to the Application Rule to this requirement, this Principle implies that due regard is given to structural safety and serviceability, including durability.

According to **C2.1(1)P** structures designed to EC7 shall be in compliance with the general design principles given in EC1-1. EC7 interprets these design principles for the case of geotechnical designs by providing guidance on the design process and on the minimum requirements for the extent and quality of geotechnical investigations, calculations, construction control checks, and monitoring and maintenance of the completed structure necessary to satisfy the fundamental requirements. The EC7 geotechnical design process is shown in Figure 2.1. This process includes assessing the complexity of the design, carrying out geotechnical

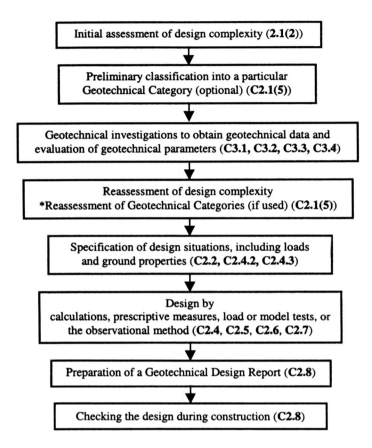

* Reassessment of the Geotechnical Categories should be carried out at each stage during the design process

Figure 2.1: *EC7 Geotechnical design process with relevant clauses shown in brackets*

investigations to obtain geotechnical data, specifying the design situation including loads, designing by calculations, prescriptive measures, load or model tests, or the observational method, and preparing a design report.

2.2.2 Geotechnical Requirements

The minimum requirements for the extent and quality of geotechnical investigations, calculations and construction control checks should take account of the complexity of a particular project and the risks to property and life [C2.1(2)P]. The complexity of geotechnical designs varies greatly from very simple, light structures on good firm ground to complex structures such as large, sensitive structures on soft ground or deep excavations close to old buildings. EC7 distinguishes between light and simple geotechnical structures, for which it is possible to ensure that the fundamental requirements are satisfied on the basis of

Reference	Factors affecting the complexity of a design	Considered
1	Nature and size of the structure	
2	Conditions with regard to the surroundings, for example neighbouring structures, utilities, traffic, etc.	
3	Ground conditions	
4	Groundwater situation	
5	Regional seismicity	
6	Influence of the environment, e.g. hydrology, surface water, subsidence, seasonal changes of moisture	

Table 2.1: *Factors affecting the complexity of a geotechnical design*

experience and qualitative geotechnical investigations with negligible risk for property, and more complex geotechnical structures, for which calculations and more extensive investigations are necessary. In the case of structures with low geotechnical complexity, simplified design procedures, for example published presumed bearing resistances, are acceptable [**C2.1(3)**].

The factors listed in **C2.1(4)P** which need to be taken into account when assessing the complexity of a geotechnical design are shown in Table 2.1. Some guidance on the significance of these factors and how they should be taken into account is given in EC7. For example, the nature and size of the structure and the ground conditions may be taken into account by adopting the system of Geotechnical Categories outlined below. The conditions with regard to the surroundings are important because of the effects that new construction can have on neighbouring structures and services and the effects that neighbouring structures can have on the new construction. This is an aspect of ground-structure interaction and is discussed further in §2.2.5.

The groundwater situation is an extremely important factor in geotechnical design. It is essential that the groundwater situation be investigated so that the pore water pressures can be assessed and the significance of the groundwater understood. Many geotechnical failures, such as landslides, base heaves in excavations and piping, are due to high pore water pressures or excessive water flows in the ground.

For countries in seismic regions, the effects of seismic loading need to be taken into account. Special requirements for the geotechnical design of structures to resist seismic loading are given in EC8, Part 5. The influence of the environment includes factors such as changes in the general groundwater level; for example lowering of the groundwater level due to the extraction of water by pumping from an underlying aquifer, or, as is now occurring in some cities such as London, rising of the groundwater level due to a reduction in pumping. Other environmental factors include subsidence due to water removal by the roots of nearby trees or subsidence due to mining activities.

2.2.3   Geotechnical Categories and Geotechnical Risk

A system of three Geotechnical Categories, referred to Geotechnical Categories 1, 2 and 3 is presented in **C2.1(5)** to take account of the different levels of complexity of a design. The use of these Categories is not a code requirement as they are presented in an Application Rule and so are optional. The advantage of the Categories is that they provide a framework for categorising the different levels of risk in geotechnical design. Geotechnical risk is a function of two factors: geotechnical hazards (i.e. dangers) and the vulnerability of the structure in relation to specific hazards. With regard to the design complexity factors listed in Table 2.1, factors 1 and 2 (the structure and its surroundings) relate to vulnerability and 3 to 6 (ground conditions, groundwater, seismicity and environment) are geotechnical hazards. The various levels of complexity of these factors in relation to the different Geotechnical Categories and the associated geotechnical risks are shown in Table 2.2. It is the geotechnical designer's responsibility to ensure that structures have sufficient safety against being vulnerable to any potential hazards.

In geotechnical designs to EC7, the distinction between the Categories lies in the degree of expertise required and in the nature and extent of the geotechnical investigations and calculations to be carried out as shown in Table 2.3. Some examples of structures in each Category are also shown in this table. It should be noted that EC7 does not provide for any variation in the values of the safety factors between the Categories [**C2.4.1(2)**]; rather the fundamental requirements in the higher Categories are satisfied by greater attention to the quality of the geotechnical investigations and the design.

The quality of a geotechnical design is influenced by the qualifications of the designer. No specific guidelines are given in EC7 with regard to the qualifications required by designers for particular Categories except that **C1.4(1)P** assumes structures are designed by appropriately qualified and experienced personnel. On this point Simpson and Driscoll (1998) state that a designer must be competent to judge that the situation is not more complex than that allowed within the Category. The authors have attempted to provide in Table 2.3 an indication of the sort of person who may be competent with regard to the different Categories. The main features of the different Categories are summarised in the following paragraphs.

Geotechnical Category 1
Geotechnical Category 1 (GC1) includes only small and relatively simple structures for which the fundamental requirements may be satisfied on the basis of experience and qualitative geotechnical investigations and where there is negligible risk for property and life. Apart from the examples listed in Table 2.3, EC7 does not, and, without local knowledge, cannot provide detailed guidance or specific requirements for GC1. These must be found elsewhere, for example in local building regulations, national standards, and textbooks. The design of GC1 structures requires someone with appropriate comparable experience as defined in **C1.5.2(1)P**.

Geotechnical Category 2
Geotechnical Category 2 (GC2) includes conventional types of structures and foundations with no abnormal risk or unusual or exceptionally difficult ground or loading conditions. Structures in GC2 require quantitative geotechnical data and analysis to ensure that the fundamental requirements will be satisfied and require a suitably qualified person, normally a civil engineer with appropriate geotechnical knowledge and experience. Routine procedures may be used for field and laboratory testing and for design and construction. EC7 is chiefly concerned with the requirements for GC2 structures.

Factors to be Considered	Geotechnical Categories		
	GC1	GC2	GC3
Geotechnical hazards	*Low*	*Moderate*	*High*
Ground conditions	Known from comparable experience to be straightforward. Not involving soft, loose or compressible soil, loose fill or sloping ground.	Ground conditions and properties can be determined from routine investigations and tests.	Unusual or exceptionally difficult ground conditions requiring non routine investigations and tests
Groundwater situation	No excavations below water table, except where experience indicates this will not cause problems	No risk of damage without prior warning to structures due to groundwater lowering or drainage. No exceptional water tightness requirements	High groundwater pressures and exceptional groundwater conditions, e.g. multi-layered strata with variable permeability
Regional seismicity	Areas with no or very low earthquake hazard	Moderate earthquake hazard where seismic design code (EC8) may be used	Areas of high earthquake hazard
Influence of the environment	Negligible risk of problems due to surface water, subsidence, hazardous chemicals, etc.	Environmental factors covered by routine design methods	Complex or difficult environmental factors requiring special design methods
Vulnerability	*Low*	*Moderate*	*High*
Nature and size of the structure and its elements	Small and relatively simple structures or construction. Insensitive structures in seismic areas	Conventional types of structures with no abnormal risks	Very large or unusual structures and structures involving abnormal risks. Very sensitive structures in seismic areas
Surroundings	Negligible risk of damage to or from neighbouring structures or services and negligible risk for life	Possible risk of damage to neighbouring structures or services due, for example, to excavations or piling	High risk of damage to neighbouring structures or services
Geotechnical risk	*Low*	*Moderate*	*High*

Table 2.2: *Geotechnical Categories related to geotechnical hazard and vulnerability levels*

Geotechnical Category 3

Geotechnical Category 3 (GC3) includes structures or parts of structures that do not fall within the limits of GC1 and GC2. GC3 includes very large or unusual structures, structures involving abnormal risks, or unusually or exceptionally difficult ground conditions or structures in highly seismic areas. GC3 structures will

	Geotechnical Categories		
	GC1	**GC2**	**GC3**
Expertise required	Person with appropriate comparable experience	Experienced qualified person	Experienced geotechnical specialist
Geotechnical investigations	Qualitative investigations including trial pits	Routine investigations involving borings, field and laboratory tests	Additional more sophisticated investigations and laboratory tests
Design procedures	Prescriptive measures and simplified design procedures, e.g. design bearing pressures based on experience or published presumed bearing pressures. Stability or deformation calculations may not be necessary.	Routine calculations for stability and deformations based on design procedures in EC7	More sophisticated analyses
Examples of structures	- Simple 1 and 2 storey structures and agricultural buildings having maximum design column load of 250kN and maximum design wall load of 100kN/m - Retaining walls and excavation supports where ground level difference does not exceed 2m - Small excavations for drainage and pipes.	Conventional: - Spread and pile foundations - Walls and other retaining structures - Bridge piers and abutments - Embankments and earthworks - Ground anchors and other support systems - Tunnels in hard, non-fractured rock	- Very large buildings - Large bridges - Deep excavations - Embankments on soft ground - Tunnels in soft or highly permeable ground

Table 2.3: *Investigations, designs and structural types related to Geotechnical Categories*

require the involvement of a specialist, such as a geotechnical engineer. The requirements in EC7 for GC2 are the minimum requirements for GC3 but, apart from the GC2 provisions, EC7 does not provide any special requirements for GC3.

Classification into a particular Category
If the Geotechnical Categories are used, preliminary classification of a structure into a particular Category is normally performed prior to any investigation or calculation being carried out [C2.1(5)]. It is important to note, however, that this classification may need to be changed during or following the investigation or design as additional information becomes available. It should also be noted that, when using this system, all parts of a structure do not have to be treated according to the highest Category. Only some parts of a structure may need to be classified in a higher Category and only these parts will need to be treated differently; for example, with regard to the level of investigation or the degree of sophistication of

the design. The procedures of the higher Category may be used to justify more economical designs, or where the designer considers them to be more appropriate.

As classification of a structure into a particular Geotechnical Category is part of the geotechnical design, this classification should be the responsibility of the person responsible for the geotechnical design, not the project leader, who may not have the necessary geotechnical knowledge or experience.

Since many geotechnical designs are composed of relatively straightforward parts and more complex parts, it may be difficult to implement the system of Geotechnical Categories in practice, particularly if one part of a design is classified in one Category and another part is classified in a different Category. Furthermore there is no legal status for the Geotechnical Categories and changing the Category during construction may be difficult contractually.

2.2.4   Limit State Design Philosophy

In the limit design method, the performance of a whole structure or a part of it is described with reference to a set of limit states beyond which the structure fails to satisfy the fundamental requirements [**C3.1(1)P** of EC1-1]. In the Eurocodes, a distinction is made between ultimate and serviceability limit states as follows:

- *Ultimate Limit States* (ULS), according to **C3.2(1)P** of EC1-1, are those situations involving safety, such as the collapse of a structure or any other type of failure, including excessive deformation in the ground prior to failure causing failure in the supported structure, or where there is a risk of danger to people or severe economic loss. Ultimate limit states have a low probability of occurrence for well designed structures.

- *Serviceability Limit States* (SLS), according to **C3.3(1)P** of EC1-1, correspond to those conditions beyond which the specified requirements of the structure or structural element are no longer met. Examples include deformations, settlements, vibrations and local damage of the structure in normal use under working loads such that it ceases to function as intended. Serviceability limit states have a higher probability of occurrence than ultimate limit states.

The basis of the limit state design philosophy outlined in EC1-1, and hence adopted in EC7, is that all the possible modes of failure for a structure shall be considered and that, for each design situation, it shall be verified that no relevant limit state is exceeded [**C2.1(6)P**], i.e. it shall be demonstrated that the exceedence of any limit state is sufficiently unlikely. In limit state design, the ultimate and serviceability limit states are each normally considered separately. In practice, however, it is often known from experience which limit state will govern the design and the avoidance of the other limit states may be verified by a control check. Landslides are probably the most common ultimate limit state and excessive foundation settlements the most common serviceability limit state encountered by geotechnical engineers.

Design Approaches

The following four design approaches that may be adopted to verify that no relevant limit state is exceeded are listed in **C2.1(7)**:

- use of calculations (§2.5);
- adoption of prescriptive measures (§2.6);
- use of experimental models and load tests (§2.7);

- use of an observational method (§2.8);.

As indicated, these approaches are described in §2.5 – §2.8, respectively. Most of EC7 is concerned with the first approach, the use of calculations. Sometimes only one of these approaches, for example calculations, will enable a design to be carried out and, in many situations, experience will determine which method is the most appropriate one to use. According to **C2.1(7)**, these approaches may be used in combination; for example, in pile design, the pile resistance may be determined from load tests and this value may then used in calculations to design the pile foundations. Indeed it is often advisable in geotechnical designs to use more than one approach. Experience, particularly of the local ground conditions and similar designs, is of particular importance when using any of the above design approaches so, whenever possible, designs shall be checked against comparable experience [**C2.1(11)P**].

2.2.5   Ground-Structure Interaction

EC7 requires that the interaction between the structure and the ground shall be considered [**C2.1(8)P**]. Ground-structure interaction, often referred to as soil-structure interaction, is the process by which the stiffness of the structure and the stiffness of the ground affect the stresses on the structure and in the ground and hence the deformation of the structure. Other types of ground-structure interaction which should be considered include the effects of new construction, such as the driving of piles, excavations and tunnels, on neighbouring structures.

In geotechnical designs, redistribution of loading in statically indeterminate structures due to ground-structure interaction, for example beneath a series of pad foundations, tends to transfer the structural loading from weaker to stronger zones of ground more capable of bearing the loading. Similarly, flexure of a sheet pile retaining wall tends to redistribute the earth pressure loading on the wall so as to minimise the bending moments in the wall. Hence in geotechnical designs, load redistribution in statically indeterminate structures due to ground-structure interaction is generally beneficial.

2.3   Design Situations

According to **C2.3(1)P** of EC1-1, design situations are the circumstances in which the structure may be required to fulfil its function. The selected design situations should be sufficiently severe and so varied as to encompass all conditions which can reasonably be foreseen to occur during the execution and use of the structure. EC1-1 classifies design situations into:

- persistent situations, which refer to the conditions of normal use;
- transient situations, which refer to temporary conditions during execution or repair;
- accidental situations, which refer to exceptional conditions such as fire, impact or explosion; and
- seismic situations, which refer to the exceptional conditions experienced during a seismic event.

Reference	Factors to be considered	Considered
1	General suitability of the ground on which the structure is located	
2	Disposition & classification of relevant zones of soil and rock	
3	Mine workings, caves and other underground works	
4	For structures resting on or near rock: • dipping bedding planes • interbedded hard and soft strata • faults, joints and fissures • solution cavities and swallow holes filled with soft material	
5	The actions and their combinations	
6	The nature of the environment including: • the effects of scour, erosion and excavation on the geometry of the ground surface • the effects of chemical corrosion • the effects of weathering • the effects of freezing	
7	Variation of groundwater levels including effects of dewatering, flooding and the failure of drainage systems	
8	The presence of gases emerging from the ground	
9	Other effects of time and the environment on the properties of materials	
10	Construction sequence (not actually included in C2.2)	
11	Earthquake hazard	
12	Subsidence due to mining and other causes	
13	The tolerance of the structure to deformations	
14	The effects of the new structure on existing structures and services	

Table 2.4: *Factors to be considered when specifying a design situation*

C2.2(1)P provides a list of factors to be considered, as appropriate, when specifying a design situation in designs to EC7. This list is presented in Table 2.4 and is similar to the list of factors in Table 2.1 which need to be considered when assessing the complexity of a geotechnical design. However the list in Table 2.4 is much more detailed and comprehensive, focusing particularly on.the factors relating to the ground and the loading on a structure that need to be considered.

2.4 Durability

EC7 requires [**C2.3(1)P**] that the environmental conditions be estimated to assess their significance in relation to the durability of the structure being designed. Depending on the results of this assessment, it may be necessary to make special provisions to protect or improve the resistance of the structural element in the ground. The concrete cover requirements for durability are given in **C4.1.3.3(108)** of EC2 - Part 3 (ENV 1992-3, 1998). **C4.1.3.3(111)** of EC2 - Part 3 recommends that a blinding layer of concrete be used under concrete elements to cover the soil to avoid contamination.

An important environmental component that needs to be considered with regard to the durability of a buried structure is the groundwater level and its variation during construction and during use of the structure. Lowering the groundwater level may have the undesirable effect of drawing oxygen down into the ground and changing the ground conditions from being anaerobic to aerobic, thus causing the ground to become more aggressive.

There is a need for care in the design of steel sheet pile walls where the ground is sufficiently permeable to allow the percolation of groundwater and oxygen to cause corrosion of the steel, and where the walls are exposed to free water, particularly in the mean water level zone. Another example of corrosion, mentioned in **C2.3(2)**, that should be considered is the pitting of steel embedded in fissured or porous concrete, particularly in the case of rolled steel where the mill scale, acting as a cathode, promotes electrolytic action with the scale-free steel surface acting as an anode. The general requirements for protecting steel piling against the effects of corrosion are given in **C2.6.1** of EC3 - Part 5 (prENV 1993-5, 1997). The specific durability requirements for steel bearing piles are given in **C2.6.2** of EC3 - Part 5 and, for sheet piling, in **C2.6.3** of EC3 - Part 5.

Timber elements, such as piles, buried in the ground, are particularly prone to aggressive attack by fungi and aerobic bacteria if the groundwater level is lowered and they are exposed to increased levels of oxygen.

Concrete foundations may be attacked by aggressive agents such as acidic conditions or sulphate salts in the groundwater. However it is generally found that most concrete foundations in natural ground are not subject to corrosion and hence special provisions are not normally required to protect concrete foundations.

There are many different types of synthetic fabrics, each offering different levels of resistance to environmental conditions. To assess the durability of these products the designer should investigate the properties of the particular fabric being considered for design. The most important properties to be considered in the case of synthetic fabrics are the ageing effects when the fabric is exposed to ultraviolet (UV) light, ozone degradation or the combined effects of temperature and stress. Another factor that should be considered is degradation due to chemical attack.

2.5   Geotechnical Design by Calculation

2.5.1   Limit State Design Calculations

Design by calculation is the most common geotechnical design approach. **C2.4.1(3)P** requires that the calculation model shall describe the behaviour of the ground for the limit state under consideration. Thus separate and different calculations should be carried out when checking for ultimate and serviceability limit states. The ULS calculation will normally involve analysing a mechanism and using ground strength properties while the SLS calculation will normally involve a deformation analysis and ground stiffness or compressibility properties. Design calculations shall be carried out in accordance with the partial factor method in EC1-1 [**C2.4.1(1)P**]. The basis of this method is that calculations are used to verify that the risk of failure of a structure, whether by an ultimate or serviceability limit state type of failure, is acceptably low.

Traditionally geotechnical calculations have involved either plastic collapse mechanisms for stability analyses or pseudo-elastic stress states for deformation analyses. However these two different types of calculations do not always correspond to the ultimate and serviceability limit states as failure in a structure can occur due to excessive movement in the ground as well as due to the formation of a mechanism in the ground. Examples of movement that may cause an ultimate limit state are excessive settlements of foundations, compression of soil causing downdrag on piles and heave in swelling ground.

Separate and different calculation models for ultimate and serviceability limit states are direct models that analyse the relevant limit state. When no reliable calculation model is available for a specific limit state, **C2.4.1(7)P** requires the analyses of other limit states with factors to ensure that this limit state is sufficiently unlikely. This is the indirect design method and its use is described in §2.5.8.

The need to ensure that, wherever possible, calculation models are correlated with field observations of previous designs, model tests or more reliable designs is noted in **C2.4.1(8)P**. According to **C2.4.1(9)**, the calculation model may consist of an empirical relationship between the test results and design requirements in place of an analytical model. However the empirical relationship needs to be clearly established for the relevant ground conditions.

The components involved in any limit state geotechnical design calculation are illustrated in Figure 2.2. These components are:

- imposed loads or displacements (referred to as actions in the Eurocodes);
- properties of soil, rock and other materials;
- geometrical data;
- partial factors or some other safety elements;
- limiting or acceptable values of deformations, crack widths, vibrations, etc.
- calculation models.

These components are inter-related; for example the way in which the loads and ground properties are defined and chosen determines what safety factor values are required. For this reason it is important to view all the design requirements in EC7 as forming the parts of a consistent system and one should beware of trying to introduce requirements from other codes into designs being carried out in accordance with EC7.

Design values
It should be noted that design values are the values of parameters used in design calculations.

2.5.2 Actions

Actions in Geotechnical Design
The term 'actions' is introduced in the Eurocodes for loads and imposed displacements. However actions are defined in **C2.4.2(1)P** more specifically for

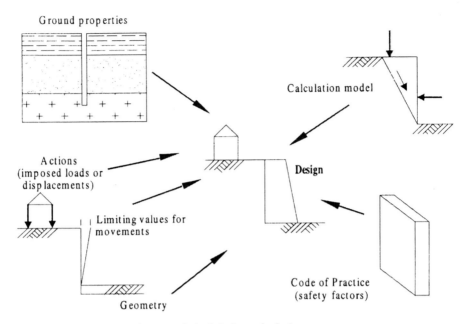

Figure 2.2: *Components of a geotechnical design calculation*

geotechnical designs as forces or imposed displacements that are known quantities in the calculation model which the designer shall choose or determine before the calculation is carried out. In geotechnical design, certain forces are sometimes actions and sometimes unknown internal forces, depending on the design situation; for example, in the case of a stem cantilever retaining wall, the earth pressure on the back of the wall is an action in the design of the wall but is an internal force (i.e. not an action) when analysing the overall stability of the wall. **C2.4.2(4)P** provides a list of items to be considered as actions which includes:

- weight of soil, rock or water;
- in situ stresses in the ground;
- free water, groundwater and seepage pressure;
- dead, imposed and environmental loads from structures;
- surcharges, imposed loads, pre-stress forces and traffic loads;
- removal of load or excavation of ground;
- movements and accelerations due to earthquakes, blasting, vibrations, etc.;
- movements due to creeping or sliding soil masses or soil self-compaction;
- swelling and shrinkage due to vegetation, climate or moisture change;
- temperature effects including frost heave and ice loading.

It should be noted that earth pressures, which may be considered as actions in some situations and resistances in other cases, are not included in this list. However, it is likely that lateral earth pressures, whose values are fixed and known at the start of a design calculation, will be listed as actions in the EN version of EC7. Passive earth pressure that resists sliding is a resistance, and so is not an action.

Permanent actions in geotechnical designs include the self weight of structural and non structural elements and also include the forces due to ground, groundwater and free water [**C2.4.2(16)P**].

In some situations the effect of ground-structure interaction can modify, i.e. redistribute, the loads; for example the bearing pressures beneath the foundations of a redundant structure depend on the relative stiffnesses of the structure and the ground. Where modification of the loads is significant, a ground-structure interaction analysis may need to be carried out to determine the modified load distribution for the design of the foundation.

Water Pressures

EC7 recognises the particular importance of water and water pressures in geotechnical design. **C2.4.2(8)P** requires that special consideration be given to the safety evaluation of geotechnical structures where water pressures are the predominant forces. This is because changes in the water table level can have a significant effect on the safety and because, as a permanent action, no partial factor is applied to the water pressure force for Case C (see Table 2.6). Factors which may change the water level and water pressures and which need to be considered include:

- deformations of the ground;
- soil fissuring and variable permeability, with the inherent risk of erosion;
- the effects of drainage;
- the supply of water by rain, flood, burst water mains, etc.;
- changes of water pressure due to growth or removal of vegetation.

Characteristic Values of Actions

The characteristic values of permanent actions, G_k that are derived from the weight of materials, including water pressures, are normally calculated using the given or nominal unit weights of the materials with no account being taken of the variability in the unit weights. Characteristic earth pressures are obtained using characteristic ground properties and surface loads and include characteristic water pressures.

The characteristic values of variable actions, Q_k, for example wind and snow loads, are either specified values or values obtained from meteorological records for the area concerned.

ULS Design Values of Actions

The design values of actions, F_d for ULS analyses shall either be derived from characteristic values, F_k using the following equation:

$$F_d = \gamma_F F_k \qquad (2.1)$$

where γ_F is the partial load factor, or else shall be assessed directly. This equation is not given in the ENV, but has been taken from **C9.3.1(P)** of EC1-1 and is likely to be included in the EN version, as is the text, which is modelled on the equivalent text in **C2.4.3(1)P** for the design values of ground properties. Where the action consists of permanent and variable loads, Equation 2.1 may be written as:

$$F_d = \gamma_G G_k + \gamma_Q Q_k \qquad (2.2)$$

The values of γ_G and γ_Q are given in Table 2.6. In the case of a statically indeterminate structure, some load redistribution may occur so that it may be necessary to obtain the F_d value required for the geotechnical analysis from a structural analysis rather than from a simple application of Equation 2.2.

Design values of actions due to ground and groundwater may be derived by methods other than by the use of partial factors. In this case the values in Table 2.6 shall be used as a guide to the required level of safety as these are considered appropriate for conventional designs [C2.4.2(19P]. Direct assessment of design values is particularly appropriate when the values derived using the partial factors in Table 2.6 are clearly impossible.

Design water pressures
The ULS design values for water pressures and seepage forces shall be the most unfavourable values that could occur in extreme circumstances [C2.4.2(10P], for example in the case of flooding or a burst water main. The need to treat the water table rising to the surface due to inadequacy of the drainage system is given as an example of an extreme circumstance [C2.4.2(11)]. According to this clause, it will often be necessary to assume the water table can rise to the surface for ULS designs unless the adequacy of the drainage system and its maintenance can be ensured. In Case B (see §2.5.6), ULS design values of groundwater pressures may be derived by applying a partial load factor to the characteristic groundwater pressures or by applying a safety margin to raise or lower the characteristic groundwater level.

2.5.3 Ground Properties
Characteristic Values of Ground Properties
The characteristic value of a ground property is defined in **C2.4.3(5)P** as a cautious estimate of the value affecting the occurrence of the limit state. Thus the characteristic value needs to be selected with reference to the particular limit state, and hence a particular parameter, e.g. ϕ' in any one stratum, may have different characteristic values for different failure mechanisms. The value affecting the occurrence of a limit state is the mean (i.e. average) value over the relevant volume of ground [**C2.4.3(6)**]. For example, in the case of a slope failure involving a large volume of ground, the mean strength value governing the stability of the slope is the average of all the strengths along the slip surface which may pass through both strong and weak zones. In contrast, the failure mechanism of a single pile foundation involves a much smaller volume of ground, and, if the pile is founded in a weak zone of the same ground as the slope, the mean value governing the stability of the pile will be lower than that governing the slope stability. Hence the characteristic value chosen to design the pile will be a cautious estimate of the local mean value and will be more cautious (lower) than the characteristic global mean value of the same parameter for the same stratum chosen to design the slope. The problem in selecting the characteristic value, X_k of a ground property is deciding how cautious this mean value should be. This problem is illustrated in Example 2.1.

Where test results are obtained from different types of tests, e.g. vane tests, SPT tests and triaxial tests, different derived values of the same parameter may be obtained at the same location, as noted in §3.3.1, and this should be taken into consideration when determining the characteristic value. It may be necessary to apply a conversion factor to the test results so that they represent the behaviour of the soil in the ground for the particular limit state [**C2.4.3(3)**].

Undrained shear strength, c_u (kPa)

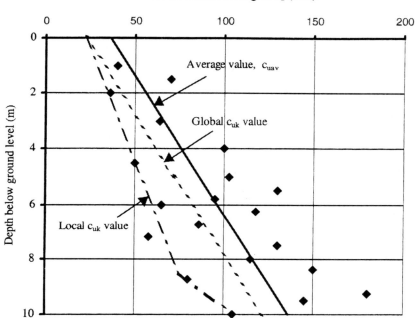

Figure 2.3: *Characteristic values for different design situations*

Which strength value - peak, critical state, mobilised or residual?
In accordance with **C2.4.3(5)P,** and as explained by Simpson and Driscoll (1998), the relevant strength value to be used in design is the one affecting the occurrence of the limit state under consideration. Although not stated explicitly in EC7, the relevant strength value for the occurrence of an ultimate limit state of collapse is generally the peak value, like the value traditionally selected for design, not the mobilised value. In just a few cases EC7 does give guidance on the relevant value; for example, in the case of the sliding of a spread foundation, **C6.5.3(3)** recommends that post-peak behaviour be considered, while in the case of a concrete or steel sheet piled wall supporting sand or gravel, **C8.5.1(4)** recommends that the mobilised friction between the ground and the wall, δ may be assumed to be $k\phi'$ where, due to disturbance, the ϕ' value of the ground at the ground wall interface should not exceed the critical state value.

Example 2.1: Selection of Global and Local Characteristic Values
The undrained shear strength, c_u values obtained from twenty tests on clay soil are plotted against depth in Figure 2.3. The global and local c_{uk} values are required.

The solid line plotted in Figure 2.3 gives the average values of the test results, c_{uav}. Taking all the test results, a cautious estimate of these average values corresponding to the global c_{uk} value for a failure mechanism passing through all this ground is plotted as the dotted line. This estimate is based on comparable experience and on the scatter of the data (the same line is obtained using Equation 2.5). If the failure mechanism only involves a small volume of the ground, as in the case of the base resistance of a pile, then the local c_{uk} value would be more cautious and correspond to the dashed line in Figure 2.3. This line represents a lower bound to the test results or the worst credible values. It can be seen from this example that the characteristic values of ground properties are similar to the values that have conventionally been selected for use in geotechnical designs, as noted by Orr (1994).

Statistical Definition of Characteristic Value
The characteristic value of a material property is defined statistically in C5(2) of EC1-1 as the 5% fractile for strength parameters and the mean value for stiffness parameters in a hypothetical unlimited test series. These definitions relate particularly to manufactured materials, such as steel and concrete, which are prepared and tested under controlled conditions. Soil and rock, as natural materials, are more variable than manufactured materials and only a very small portion of the total volume involved in a design situation is tested. Consequently the use of statistics to assess material properties is not as applicable in geotechnical design as in designs involving manufactured materials.

If statistics are used, the characteristic value of a ground property is defined in C2.4.3(6) as the value such that the probability of a worst value governing the occurrence of a limit state is not greater than 5%. This provides a target value for selecting the characteristic value and applies to the ground stiffness as well as the strength parameters, as discussed in §2.5.7. It should be noted that this statistical definition of the characteristic value relates to the mean value of the ground property in situ, which is the value governing the occurrence of the limit state, and not the 5% fractile of a series of derived values obtained from test results. The characteristic value corresponds to a 95% confidence level that the actual mean value, X_m is greater than the selected characteristic value.

If statistical methods are used to determine the characteristic value, they should be used with caution and should take account of comparable experience [C2.4.3(6)]. Inappropriate characteristic values may be obtained if a pure statistical approach is adopted, ignoring the actual design situation and comparable experience.

Statistical Methods for Determining the Characteristic Value
Using statistics, the characteristic value, X_k of a soil property, corresponding to a 95% confidence level that the actual mean value, X_m is greater than this value, is given by:

$$X_k = X_m[1 - k_n V] \qquad (2.3)$$

where k_n is a factor, depending on the type of statistical distribution and the number of test results, and V is the coefficient of variation (standard deviation (σ)/mean value). Since the actual mean value, X_m of a soil parameter cannot normally be determined statistically by a sufficient number of tests, it must be assessed from the average value, X_{av} of the test results.

W.S. Gossett, who was employed by the Guinness Brewery, Dublin in the early part of the twentieth century, published tables of statistical values for assessing, with different levels of confidence, the mean density of beer in large vats from the

Soil property	Range of typical V values	Recommended V value if limited test results available
tan ϕ'	0.05 - 0.15	0.10
c'	0.30 – 0.50	0.40
c_u	0.20 – 0.40	0.30
m_v	0.20 – 0.70	0.40
γ (unit weight)	0.01 – 0.10	0

Table 2.5: *Typical values for V, the coefficient of variation*

results of tests on a limited number of samples. As Gossett was not permitted to publish his values under his own name, he published them under the pseudonym of Student (1908). His t values are introduced into Equation 2.3 as follows:

$$X_k = X_m[1 - \frac{tV}{\sqrt{n}}] = X_m - \frac{t\sigma}{\sqrt{n}} \qquad (2.4)$$

where t is known as the Student value and depends on the number of test results, n and the level of confidence required, and the other parameters are as defined above. If the ground is considered to be homogeneous, then the Student t value for a 95% confidence level may be used to determine the X_k value. Some typical values of V for different soil properties, based on Schneider (1997), are given in Table 2.5.

The problem with using Student t values to select the characteristic value of geotechnical parameters, is that the number of test results is usually very limited, normally less than 10. As the number of test results, n decreases, the value of t increases at an increasing rate so that the ratio $t/\sqrt{n}$ increases even more. Consequently, using this purely statistical method without taking account of experience of the ground conditions, the value obtained for X_k is usually too low, i.e. too cautious, and uneconomic.

Schneider's Method for Determining the Characteristic Value
Schneider (1997), on the basis of comparative calculations, has shown that a good approximation to X_k is obtained when $k_n = 0.5$; i.e. if the characteristic value is chosen as one half a standard deviation below the mean value, as in the following equation:

$$X_k = X_m[1 - k_n V] = X_m - 0.5\sigma \qquad (2.5)$$

Example 2.2: Use of Statistics to Determine Characteristic Value

The characteristic value of the angle of shearing resistance, ϕ'_k is required for a 10m depth of ground consisting of sand for which the following ϕ' values, plotted in Figure 2.4, were determined from 10 triaxial tests: 33.0°, 35.0°, 33.5°, 32.5°, 37.5°, 34.5°, 36.0°, 31.5°, 37.0°, 33.5°.

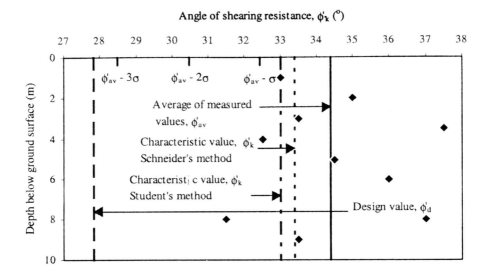

Figure 2.4: *Determination of ϕ'_k value for homogeneous ground*

Design Situation
No consistent variation in the ϕ' value was found with depth, so the sand stratum is assumed to be homogeneous.

Calculations
The average angle of shearing resistance from these test results, ϕ'_{av} is 34.40° with standard deviation, $\sigma = 1.97°$ and coefficient of variation, $V = 0.057$. The Student t value for a 95% confidence level with 10 test results is 2.26 so that, from Equation 2.4 and assuming ϕ'_{av} equals ϕ'_m, the ϕ'_k value is $34.4 - 1.97 * 2.26/\sqrt{10}) = 33.0°$. As the number of test results decreases, the ϕ'_k value obtained using this method becomes more cautious. Using Schneider's method and Equation 2.6, ϕ'_k is $34.4 - 0.5 * 1.97 = 33.4°$, which is less cautious than the value obtained using the Student t value. These values are plotted in Figure 2.4. For comparison, the values of one, two and three standard deviations (σ, 2σ and 3σ) below the average value are also plotted in Figure 2.4.

Design Values of Ground Properties
The ULS design value of a ground property, X_d shall either be obtained by dividing X_k by the partial factor for the material property, γ_m as follows:

$$X_d = X_k / \gamma_m \qquad (2.6)$$

or shall be assessed directly [**C2.4.3(1)P**]. Values for γ_m for the design cases described in §2.5.6 are given in Table 2.6.

When X_d is assessed directly, not using γ_m factors, **C2.4.3(14)P** requires that the γ_m values given in Table 2.6 be used as guidance to the required level of safety.

Example 2.3: Determination of the Design Value

The Case C design value of the angle of shearing resistance, ϕ'_C is required for the 10m depth of ground consisting of sand in Example 2.2.

Applying the γ_m value of 1.25 for Case C in Table 2.6 to the ϕ'_k value of 33.4° in the Example 2.2 gives ϕ'_C = atan(tanϕ'_k/1.25) = 27.8°. This design value is plotted in Figure 2.4 and is found to be just over three standard deviations below, ϕ'_{av} the average angle of shearing resistance of the test results.

2.5.4 Geometrical Data

The geometrical data involved in geotechnical design calculations include:

- the level and slope of the ground surface;
- water levels;
- the level of interfaces between strata;
- excavation levels;
- the shape of the foundation or other structural elements.

Characteristic Values of Geometric Data
The measured ground surface level, the observed or deduced strata boundaries and groundwater levels and specified excavation levels are normally chosen as the characteristic geometrical data for use geotechnical designs to EC7.

Design Values of Geometrical Data
For limit states with severe consequences, that is for ultimate limit states, the design values of geometrical data shall be the most unfavourable values that could occur in practice [C2.4.5(3)P]. The example referred to in §2.5.2 of the water table rising to the surface is given as an example of a most unfavourable, i.e. an extreme, value.

Where variations in the geometrical data are not important, C2.4.5(2)P requires that they be accounted for in the selection of the design values for the material properties or the loads. The partial load and material factors, γ_F and γ_m, include an allowance for minor variations in geometrical data. However, where the variations in the geometrical data are important, they should not be allowed for in the partial load and material factors but rather should either be assessed directly, by adopting more severe design geometrical data a_d for the problem, or be derived from the characteristic geometric data, a_k using the following equation:

$$a_d = a_d \pm \Delta_a \qquad (2.7)$$

where Δ_a is a geometric safely margin.

An example where the variation of geometrical data is important is in the design of an embedded retaining wall covered in Section 8 of EC7. C8.3.2.1(2) recommends that, in ULS calculations, the design ground surface in front of a wall whose stability depends on the passive pressure should be obtained by lowering the characteristic ground surface by an amount Δ_a.

2.5.5 Structural Properties

The design strength properties of structural materials and the design resistance of structural elements for use in calculations involving the ground, for example in the design of concrete in pad foundations or steel in sheet pile walls or ground anchors, shall be determined in accordance with the relevant Eurocodes for the particular material [C2.4.4(1)P].

2.5.6 Ultimate Limit State Calculations

In a ULS calculation involving failure in the ground, the design situation corresponding to an acceptably low risk of failure is normally achieved by applying two sets of partial safety factors, one set to increase the loads or load effects to obtain the ULS design action effect, E_d and another set to reduce the ground strength parameters or resistance to obtain the ULS design resistance R_d. To verify that the occurrence of a ULS is sufficiently unlikely, it is necessary to satisfy the following condition:

$$E_d \leq R_d \qquad (2.8)$$

The partial factors to obtain E_d and R_d may be introduced using either of the following two approaches:

- *Material Factor Approach* (MFA), in which partial factors are applied to characteristic actions and characteristic material properties, or
- *Resistance Factor Approach* (RFA) in which the partial factors are applied to characteristic action effects and characteristic resistances.

The terms MFA and RFA are not in the ENV version of EC7 but may be included in the EN version. The RFA method may appear very similar to the Load Resistance Factor Design (LRFD) method adopted for limit state design in North America (Duncan et al., 1989). However, the terms RFA and MFA were selected for use in EC7 to indicate that the RFA method is not the LRFD method but is a method fully within the partial factor framework as defined in EC1-1.

ULS Design Effect of Actions
When using the MFA method defined above, the ULS design action effect is derived from the following equation:

$$E_d = E(F_d, X_d, a_d) = E(\gamma_F F_k, X_k/\gamma_m, a_k \pm \Delta_a) \qquad (2.9)$$

where E() is the action effect function combining the design (i.e. factored) values of the actions, material properties and geometrical data. Where the effect of geometrical variations are not important, they may, as noted in §2.5.4, be allowed for in the γ_F and γ_m values, in which case $\Delta_a = 0$ and E_d simplifies to:

$$E_d = E(\gamma_F F_k, X_k/\gamma_m, a_k) \qquad (2.10)$$

When using the RFA method, the ULS design action effect is derived from the following equation:

$$E_d = \gamma_E E(F_k, X_k, a_k) \qquad (2.11)$$

where γ_E is the partial action effect factor and the actions, material properties and geometric data all have their characteristic values.

The terms X_k and X_d, representing material properties, appear in the action effect Equations 2.9 to 2.11, because the actions on some geotechnical structures, e.g. the earth pressure on a retaining wall, depend on the strength of the ground, as in Example 2.4.

ULS Design Resistance
When using the MFA method, the ULS design resistance is derived from the following equation:

$$R_d = R(F_d, X_d, a_d) = R(\gamma_F F_k, X_k/\gamma_m, a_k \pm \Delta_a) \qquad (2.12)$$

where $R()$ is the resistance effect function combining the design values of the actions, material properties and geometrical data. Where the effect of geometrical variations are not important, they may, as in the case of E_d, be allowed for in the γ_F and γ_m values, in which case $\Delta_a = 0$ and R_d simplifies to:

$$R_d = R(\gamma_F F_k, X_k/\gamma_m, a_k) \qquad (2.13)$$

When using the resistance factor approach, the ULS design effect of actions is derived from the following equation:

$$R_d = R(F_k, X_k, a_k) / \gamma_R \qquad (2.14)$$

where γ_R is the partial resistance factor and the actions, material properties and geometric data all have their characteristic values.

The terms F_k and F_d, representing actions, appear in the resistance Equations 2.12 to 2.14, because the resistances for many geotechnical structures, e.g. the bearing and sliding resistances of spread foundations, depend on the magnitude of the applied loading.

Design Cases and Partial Factors
In structural design, various load cases are normally considered corresponding to situations where one load is at its factored, i.e. most pessimistic, value while the other loads are at their characteristic (unfactored) or partially factored values and so are less pessimistic. In geotechnical designs to EC7, three (design) cases, known as Cases A, B and C, have been introduced to ensure that the risk of failure either in the structure or the ground is acceptable low for different factored combinations of loads and ground properties [**C2.4.2(13)**].

The values for the partial load and material factors presented in Table 2.1 of EC7 are shown in Table 2.6. These values are for conventional structures in persistent and transient situations. Higher values may be considered in cases of abnormal risk or unusual or exceptionally difficult ground or loading conditions. According to [**C2.4.2(14)P**], less severe values may be used in the design of temporary structures where the risk can be justified on the basis of the possible consequences. For accidental situations, all partial load factors are equal to unity.

As it is intended that the Eurocodes shall form a harmonised set of standards for structural design, the γ_F values for Cases A, B and C for permanent and variable actions in Table 2.1 of EC7 for geotechnical design are the same as the values in Table 9.2 of EC1-1 which are applicable for designs involving all materials ranging from concrete and steel to soil. The additional Cases C2 and C3 in Table 2.6 are discussed below.

C2.4.2(12)P requires that designs be verified, with regard to both the structural

Parameter	Factor	Case A	Case B	Case C	Case C2	Case C3
Partial load factors (γ_F)						
Permanent unfavourable action	γ_G	**1.00**	**1.35**	**1.00**	*1.35*	*1.00*
Variable unfavourable action	γ_Q	**1.50**	**1.50**	**1.30**	*1.50*	*1.20*
Permanent favourable action	γ_G	**0.95** **(0.90)**	**1.00**	**1.00**	*1.00*	*1.00*
Variable favourable action	γ_Q	**0**	**0**	**0**	*0*	*0*
Accidental action	γ_A	**1.00**	**1.00**	**1.00**	*1.00*	*1.00*
Partial material factors (γ_m)						
tanϕ'	$\gamma_{tan\phi'}$	**1.10**	**1.00**	**1.25**	*1.00*	*1.20*
Effective cohesion c'	$\gamma_{c'}$	**1.30**	**1.00**	**1.60** *(1.25)*	*1.00*	*1.20*
Undrained shear strength c_u	γ_{cu}	**1.20**	**1.00**	**1.40**	*1.00*	*1.40*
Compressive strength q_u	γ_{qu}	**1.20**	**1.00**	**1.40**	*1.00*	*1.40*
Pressuremeter limit pressure p_{lim}	γ_{plim}	*1.40*	*1.00*	*1.40*	*1.00*	*1.40*
CPT resistance	γ_{CPT}	*1.40*	*1.00*	*1.40*	*1.00*	*1.40*
Unit weight of ground γ	γ_g	*1.00*	*1.00*	*1.00*	*1.00*	*1.00*
Partial resistance factors (γ_R)						
Bearing resistance	γ_{Rv}	-*	**1.00**	**1.00**	*1.40*	*1.00*
Sliding resistance	γ_{Rs}	-*	**1.00**	**1.00**	*1.10*	*1.00*
Earth resistance	γ_{Re}	-*	**1.00**	**1.00**	*1.40*	*1.00*
Pile base resistance	γ_b	-*	**1.00**	**1.30**	*1.30*	*1.00*
Pile shaft resistance	γ_s	-*	**1.00**	**1.30**	*1.30*	*1.00*
Total pile resistance	γ_t	-*	**1.00**	**1.30**	*1.30*	*1.00*
Pile tensile resistance	γ_{st}	*1.40*	**1.00**	**1.60**	*1.40*	*1.00*
Anchor pull-out resistance	γ_A	*1.30*	**1.00**	**1.50**	*1.20*	*1.00*
Partial action effect and resistance model factors ($\gamma_E, \gamma_{sd}, \gamma_{rd}$)						
Action effects and resistances	γ_{sd}, γ_{rd}	*1.00*	*1.00***	*1.00*	*1.00*	*1.40*

Values in **bold** are partial factors either given or implied in the ENV version of EC7.
Values in *italics* are proposed partial factors not in the ENV that may be in the EN version.
* Partial factors that are not relevant for Case A.
** The use of a model factor greater than unity is discussed in §8.2.

Table 2.6: *Partial factors for ultimate limit states in persistent and transient situations*

and geotechnical aspects, for each of the three Cases, A, B and C, separately. Often, however, only one calculation will be necessary as it will usually be clear which case is critical in each design situation.

Boxed Values
The partial factors in Table 2.6 and the other safety elements in EC7 are presented as boxed values, i.e. these are indicative values which are identified by [ ]. It was originally intended that each CEN Member State or its Standards Organisation

would publish a National Application Document (NAD) for EC7 with the values to be used in these countries. However, as the Eurocodes have developed and matured, it has become clear that most of the partial factors and other safety elements will be fixed and not presented as boxed values when the Eurocodes, including EC7, are published as European Standards (ENs).

Purposes of Cases A, B and C
- *Case A* deals primarily with uncertainties in unfavourable variable actions and favourable permanent actions, in situations where the strengths of the structure and the ground are insignificant. Case A, aims to provide safe geotechnical sizing and structural design against problems involving gross displacement, such as buoyancy, hydraulic failure, overturning of structures, etc. Case A, therefore, is relevant in situations where equilibrium depends primarily on the weight, with little contribution from the soil strength, and where hydrostatic forces are often the main loads. Example 6.5 is a design situation involving buoyancy where Case A controls the design.
- *Case B* deals primarily with uncertainties in actions and hence the partial factors on actions for Case B are generally greater than 1.0 while the ground properties are not factored. Case B aims to provide safe geotechnical sizing and structural design against unfavourable deviations of the actions from their characteristic values, while the ground properties are equal to their characteristic values. Case B is generally an MFA method, as defined after Equation 2.8, as the individual actions are factored. Case B is usually critical in the structural design of elements such as foundations and retaining walls. Case B is not relevant where no structural strength is involved, e.g. in slope design.
- *Case C* deals primarily with uncertainties in material properties and hence is generally an MFA method, with the partial factors on ground properties greater than 1.0. In the case of piles and anchorages, however, the material factors in Case C are equal to unity while the resistance factors are greater than unity so this is an RFA method with regard to the design of piles and anchorages.

  Case C aims to provide safe geotechnical sizing and structural design against unfavourable deviations of the ground properties or resistances from their characteristic values, while the permanent actions are equal to their characteristic values and the variable actions are increased, but less so than in Case B, above their characteristic values. Case C is usually critical in determining the size of elements in the ground, such as the size of foundations and the depth of embedded retaining walls. For situations where only the strength of ground is involved, as in slope stability problems, Case C is only relevant. However, in some situations, for example a highly eccentric load on a foundation and a retaining wall supporting water rather than soil, Case B rather than Case C may be the relevant case for determining the size of the element.

Having determined the design size of a structural element from the maximum of Cases B and C, the design strength, for example the thickness a foundation or the section modulus of a retaining wall, then needs to be determined, again checking for Cases B and C. If the Case C size is used to determine the design strength in a Case B analysis, the mobilised soil strength will be less than the ultimate failure value used to determine the member size. In a retaining wall analysis, it may be found more convenient in practice to calculate both the wall length and the wall section for Cases B and C separately and then to choose the greater wall lengths and wall sections. The implications of adopting this approach are discussed in Chapter 8.

Cases C2 and C3

As noted above, except in the design of piles and anchorages, for which resistance factors are given as shown in Table 2.6, Cases B and C are MFA methods regarding resistances. This means that there is no provision in the ENV version of EC7 for the design of spread foundations or retaining structures using the RFA method with resistance factors.

To facilitate the general use of resistance factors in geotechnical designs to EC7, a Case C2 may be introduced in the EN version of EC7 as an alternative to Cases B and C. The partial factors values in Table 2.6 show that, for Case C2, $\gamma_R > 1.0$ and $\gamma_m = 1.0$ and hence it is an RFA method. It has also been proposed to introduce a Case C3 as another alternative to Cases B and C. The partial factors in Table 2.6 show that, for Case C3, $\gamma_m > 1.0$ and $\gamma_R = 1.0$ so that it is an MFA method. While the γ_m values for Case C3 are the same as for Case C, the actions are factored differently and a resistance model factor, γ_{rd} is introduced. Since Cases C2 and C3 are not included in the ENV version of EC7, no examples have been provided to show the using these cases, except for Example 2.4 on sliding, and these cases are not discussed in the subsequent chapters.

Values of Partial Factors for Cases A, B and C

The γ_F and γ_m values shown bold in Table 2.6 for Cases A, B and C are the same as those in Table 2.1 of EC7. It should be noted, however, that the γ_Q value of 0.95 for the partial factor on permanent favourable actions for Case A has been considered to be too close to unity and so is likely to be changed to 0.90 in the EN version. This new value is shown in brackets in Table 2.6. The γ_m value of 1.6 on c' may be changed to 1.25 to be the same as γ_m on $\tan\phi'_k$ and this value is also shown in brackets in Table 2.6. It should be noted that the γ_m value of 1.25 for Case C is applied to $\tan\phi'_k$, not ϕ'_k to obtain ϕ'_d for Case C. This is equivalent to reducing ϕ'_k by 3.8° for $\phi'_k = 20°$ and by 6.1° for $\phi'_k = 40°$, as shown by the graphs in Figure 2.5.

Table 2.6 has been expanded, compared with Table 2.1 of EC7, to include partial factors for the pressuremeter limit pressure, CPT resistance and unit weight of the ground. In addition, partial resistance factors have been included. These γ_R values are the partial resistance factors for driven piles given in Table 7.2 of EC7. When determining the design pile resistance, either the γ_m or γ_R factors are used, but not both. The reason for including these γ_R factors in Table 2.6 is to show that they are for Case C and that γ_R values of unity are implied in EC7 for Case B.

It should be noted that, in Table 2.6, the partial factor, γ_g on the unit weight of soil is 1.0 for all cases. This is because, in geotechnical designs, soil weight may both be a load and may also contribute to the resistance. Hence choosing a value other than unity for γ_g could result in the unrealistic situation of flat ground being found to be unstable and in the unsatisfactory situation in a stability analysis of the same factor being applied on both the loading and resistance sides of the equilibrium equation.

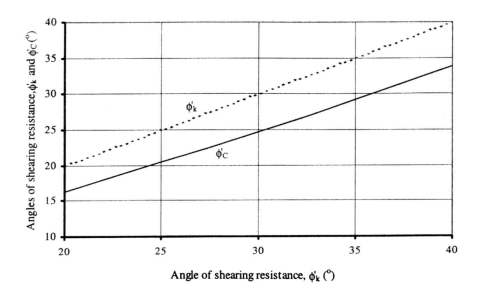

Figure 2.5: *Comparison between the characteristic and Case C design angles of shearing resistance*

Model factors

The possible use of a model factor, γ_{sd} is mentioned in **C2.4.2(15)** in connection with the design of structural elements, such as footings, piles and retaining, to increase the design action effect, if this is necessary, but no values for γ_{sd} are given in EC7. The use of model factors is also mentioned in **C2.4.2(17)** in connection with earth pressures. This clause allows the application of model factors to action effects rather than to characteristic earth pressures (i.e. for Case B calculations) to obtain the design action effect. In this situation the partial load factors are used as the model factors, as discussed in §8.2.

Single source

According to **C2.4.2(17)**, the design earth pressures for Case B are obtained by multiplying the characteristic earth pressures by the partial factors in Table 2.6. All permanent characteristic earth pressures (that is including the water pressures) on both sides of a retaining wall are treated as being from a single source and multiplied by 1.35 when the resulting action effect is unfavourable and by 1.00 when the resultant is favourable. Although **C2.4.2(17)** states that the term 'single source' is defined in EC1-1, no definition is provided there. What exactly is meant by the term 'single source' has been a point of some debate. **C2.4.2(17)** makes it clear that a single source in the case of earth pressures includes all the permanent earth and water pressure on both sides of a wall.

It appears that the concept of permanent actions from a single source also needs to be applied to the vertical loads and water pressures on a foundation in Case B as otherwise unrealistic situations can arise. For example, in the case of a spread foundation below the groundwater table, the uplift pressure on the base of the foundation, although a favourable action, should not be factored by 1.0 while the downward unfavourable water pressure on top of the foundation is factored by 1.35. Rather the resultant characteristic action effect obtained by combining all the permanent loads and water pressures on the foundation should be factored by 1.35 if unfavourable, as in Example 6.1, or by 1.0 if favourable.

Another example is a structure subjected to a large buoyancy force, as in Example 6.5. If the unfavourable water pressure in this case were factored by 1.35 and the favourable weight of the structure factored by 1.0, an unrealistic design situation with too large an uplift force would be obtained. Instead the net characteristic vertical force should calculated, again treating all the vertical forces as coming from a single source, and this net force should be multiplied by 1.35, as in Example 6.5.

Hydraulic failure

Hydraulic failure is where excessive hydraulic gradients and seepage pressures cause failure due to heave (hydraulic uplift), piping or internal erosion. EC7 provides very little information for designing against ultimate limit states involving hydraulic failure, except in **C9.5.3**, where some guidance is provided in relation to the measures to be taken to ensure the stability of slopes against internal erosion and hydraulic uplift. To remedy this omission, it is likely that the EN version will have a separate section devoted entirely to the important aspect of hydraulic failure.

To check against hydraulic uplift, it necessary to verify that at all relevant horizontal profiles the design upwards seepage pressure, u_d does not exceed the design total vertical stress, σ_d; i.e.

$$u_d \leq \sigma_d$$

where:

$u_d = h_k \gamma_w \gamma_s$ (characteristic hydraulic head at the relevant level * unit weight of water * relevant partial factor for seepage); and

σ_d = total vertical stress due to the permanent weight of soil, water and surcharge above the point being considered, divided by the partial factor for permanent favourable loads.

As hydraulic failure is a situation where it is assumed no strength of the ground is involved, Case A is the relevant case. For Case A the factor for the unfavourable action, the seepage pressure, γ_s is 1.00 as, according to **C2.4.2(16)P**, water pressure is a permanent action, while the factor on the favourable action, the total vertical stress, γ_G is 0.9, hence the overall FOS using the partial factors given in EC7 is 1.11. This design calculation relates to hydraulic heave of the ground. It is clearly important to identify the horizontal profile with the most critical seepage pressure.

Apart from providing an adequate safety margin against failure due to hydraulic heave, it is also important to take into account all possible unfavourable ground conditions. Examples include ground with thin, less permeable layers, heterogeneous ground conditions and loose medium and coarse sands and soft clays. The possibility of the interface between a construction surface and the ground becoming a seepage path and causing failure by piping needs to be considered.

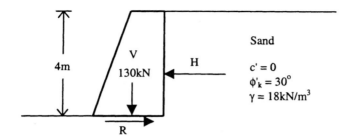

Figure 2.6: *Sliding of a gravity retaining wall*

Example 2.4: Materials Factor Approach (MFA) and Resistance Factor Approach (RFA) Analyses of a Retaining Wall

The gravity wall shown in Figure 2.6 is 4m high, weighs 130kN and supports dry sand with $c'_k = 0$, $\phi'_k = 30°$ and $\gamma = 18$ kN/m³. There is no uncertainty in the geometry so that $\Delta_a = 0$. The stability of the wall with regard to sliding is to be analysed using the MFA and RFA methods so as to clarify Equations 2.9 to 2.14.

Design Actions
For sliding, the design action is the design horizontal earth pressure force, H_d. The H_d values are calculated by the MFA and RFA methods assuming there is no friction on the back of the wall and the design active earth pressure forces are given by $(1 - \sin\phi'_d)/(1 + \sin\phi'_d)$ where the ϕ'_d values are obtained by applying the relevant partial factors given in Table 2.6 for Cases C and C2 to ϕ'_k. As γ_m for $\tan\phi'_k$ is 1.25 for Case C and 1.0 for Case C2, $\phi'_C = 24.8°$ and $\phi'_{C2} = \phi'_k = 30.0°$. Hence the H_d values using the MFA and RFA methods are:

MFA (Eqn. 2.10): $H_C = \dfrac{1}{2}\gamma h^2 K_{aC} = \dfrac{1}{2}18*4^2\dfrac{(1-\sin 24.8°)}{(1+\sin 24.8°)} = 58.9kN$

RFA (Eqn. 2.11): $H_{C2} = \gamma_E \dfrac{1}{2}\gamma h^2 K_{aC2} = 1.35*\dfrac{1}{2}*18*4^2\dfrac{(1-\sin 30°)}{(1+\sin 30°)} = 64.8kN$

Design Resistances
The design horizontal resistance, R_d is calculated using Equation 6.4, $R_d = V'\tan\delta'$, where V' is the vertical effective load normal to the wall foundation and δ' is the friction angle on the base of the wall, assumed equal to ϕ'. The R_d values using the MFA and RFA methods are:

MFA (Eqn. 2.13): $R_C = V_C\tan\phi'_C = V_k\gamma_G \tan \phi'_C = 130*1.0*\tan 24.8° = \underline{60.1kN}$

RFA (Eqn. 2.14): $R_{C2} = \gamma_R (V_k\tan \phi'_k) = 1.1*(130*\tan 30°) = \underline{82.6kN}$

Conclusion
Examining these H_d and R_d values it is seen that for both the MFA and RFA methods, i.e. for both Cases C and C2, H_d is less than R_d (58.9 < 60.1 and 64.8 < 82.6) so that Inequality 2.8 is satisfied and there is sufficient safety against sliding failure.

2.5.7 Serviceability Limit State Calculations

Calculations to assess serviceability limit states, for example the settlement of a foundation, are normally carried out at working stresses using design values of the loads and the soil deformation properties equal to the characteristic values; i.e. the partial factors on the loads and ground deformation properties are equal to 1.0 [C2.4.2(18)]. Values of the partial factors on the ground properties for the SLS may be greater than 1.0 if the properties of the ground may change during the lifetime of the structure, e.g. due to groundwater lowering or desiccation. The SLS condition to be satisfied is that the design value of the action effect, E_d (e.g. settlement), is less than the limiting value of the deformation of the structure at the serviceability limit state, C_d; i.e.

$$E_d \leq C_d \qquad (2.15)$$

Since the partial factors are equal to unity for the SLS, ensuring that the risk of the deformations exceeding the limiting values is acceptably low must be achieved through adopting sufficiently cautious values for the relevant ground properties. The characteristic values of deformation properties for SLS calculations should be selected in the same way as characteristic strength property values; i.e. they should be selected as cautious estimates of the mean values controlling the deformation of the structure with the aim that the probability of worse values governing the deformation is not greater than 5%. Choosing the mean stiffness values, which is common in structural design and stated in EC1-1, would imply that there is a 50% chance that the deformations are greater than the allowable values, which is clearly unacceptable. It is anticipated that the text of EC1-1 will be revised in this regard.

2.5.8 Indirect Design Method

Where no reliable calculation model is available for a specific limit state, analysis of the other limit states shall be carried out using factors to ensure that this limit state is sufficiently improbable [C2.4.1(7)P]. Although the ENV version of EC7 does not provide any application rules for this indirect method or any examples of its use in practice, it is expected that this method will be included in the EN version of EC7. An example where this indirect method might be used is in the design of a spread foundation, where a sufficiently high factor may be selected in a ULS type of analysis to ensure that the SLS requirements are satisfied (see §6.6.2). This method is essentially the traditional method that has been used in most countries to design foundations. It should be noted that, in this method, calculations are carried out at characteristic stress levels with unfactored load and ground parameters. One factor of safety, equivalent to a total or global factor of safety, is used both to ensure that failure does not occur and to limit deformations to acceptable values.

2.5.9 Limiting Values for Movements

Limiting values of movements are the values when an ultimate or a serviceability limit state is deemed to occur in the supported structure. The magnitude of these limiting values depend on the following factors:

• the confidence with which the acceptable limiting values can be specified;
• the type of structure;
• the type of construction material;

- the type of foundation;
- the type of ground;
- the mode of deformation;
- the proposed use of the structure.

In the design of foundations, the movements that should be considered include the settlement, relative (i.e. differential) settlement, rotation, relative rotation, tilt, horizontal displacement and vibration. **C2.4.6(4)P** requires that the design values of these movements be agreed with the designer of the supported structure. The following relative rotations are given as guidance:

1/2000 to 1/300	range of maximum relative rotations to prevent the occurrence of a serviceability limit state in different types of structure;
1/500	maximum acceptable relative rotation to prevent damage in many structures, i.e. design value for relative rotation;
1/150	relative rotation likely to cause an ultimate limit state.

According to **C2.4.6(7)**, maximum total settlements of 50 mm and maximum differential settlements of 20 mm are often acceptable, provided the relative rotations are acceptable and the settlements do not cause problems with services entering the building. When selecting acceptable movements for a structure, it is important to take account of the mode of deformation and the type of structure. For example, a brick building will be much more susceptible to hogging than sagging deformation. This may be assessed by examining the deflection ratio defined by Burland and Wroth (1975) as the ratio of the maximum displacement of the building, Δ with respect to a straight line joining two reference points on the building. If the reference points are the ends of the building, then this ratio is Δ/L, where L is the length of the building. For a masonry building that is sagging, the maximum acceptable Δ/L value is about 1/2500 while for a building that is hogging, the maximum acceptable Δ/L value is reduced to about 1/5000.

2.6 Prescriptive Measures

In certain situations calculation methods are not available or are not necessary to design against a particular mode of failure or limit state. In these situations EC7 allows the use of prescriptive measures to ensure that the occurrence of a limit states is avoided [**C2.5(1)P**]. Prescriptive measures usually consist of conventional and generally conservative details of design with attention being focused on the quality of the materials used, the workmanship and the maintenance procedures. Prescriptive measures may also be used where comparable experience, as defined in **C1.5.2(1)P**, makes calculations unnecessary [**C2.5(2)**]. Prescriptive measures generally fall into Geotechnical Category 1 designs, as indicated in Table 2.3 and do not involve analytical calculations.

Examples of prescriptive measures include selection of the depth of a foundation to avoid damage due to frost action and the planting of vegetation on slopes to prevent erosion. The design of foundations based on published presumed bearing pressures or minimum foundation widths for standard buildings on common soil types, and on comparable experience of the performance of these foundations for the relevant conditions, is another example of the use of prescriptive measures.

2.7   Experimental Models and Load Tests

Experimental models and load tests are often used in geotechnical designs to verify calculations which may be unproven, and in situations where the properties or behaviour of the ground are not adequately known or where the consequences of failure are particularly serious. Examples of situations where experimental models and load tests are used include foundations for offshore structures, pile foundations and ground anchors.

Model tests

Model tests are usually carried out on a scaled model that is much smaller than the actual construction. Great care is required in analysing the results of such an experimental model test because of possible differences between the behaviour of the soil in the model and the ground in the actual construction. One of the most important factors that may cause differences in behaviour is self weight, although this factor may be overcome by carrying out the model test in a centrifuge to obtain an increased gravity force. Other factors that may lead to differences in behaviour between the model and the actual construction and need to be considered and allowed for when using the results of model tests for design **C2.6(1)P** include:

- different ground conditions in the test;
- time effects, particularly if the duration of the test is much less than the duration of the loading in the actual construction;
- scale effects, particularly if small models are used. The effects of stress level and particle size need to be considered.

Load tests

Load tests on a particular element or part of a structure are often carried out to verify a design. Common examples of load tests include load tests on pile foundations and ground anchors. Load tests may be used both to assess the suitability of piles or ground anchors by measuring the ground resistance, and to verify the resistance assumed in the design. Normally differences in ground conditions and scale effects are of minor importance in the case of load tests. However time effects may be important, particularly in cohesive soil if the load is carried out too rapidly to allow dissipation of excess pore water pressures. In this situation the load test result will not represent the long-term behaviour and this needs to be taken into account. Example 7.2 is an example of the use load tests to design a pile. Like designs based on material properties, designs of piles and anchors based on resistances obtained from load tests should also be checked for Cases A, B and C.

2.8   The Observational Method

Since it is often difficult to predict how the ground will behave, it is sometimes appropriate to adopt the design approach known as the Observational Method [**C2.7(1)P**]. In this method the design is reviewed by monitoring the actual behaviour of the structure as it is constructed. However it is important to emphasise that the use of the observational method is more than just monitoring the behaviour during construction. It also requires some assessment of the behaviour of the structure before construction commences and the preparation of alternative designs and contingency plans should different behaviour be observed [**C2.7(2)P**]. Peck (1969), in his Rankine lecture, listed the following eight ingredients of the observational method:

1. Investigation sufficient to establish the general nature and properties of the ground;
2. Assessment of the most probable ground conditions and the most unfavourable conceivable deviations from these conditions;
3. Preparation of a working design based on the most probable conditions;
4. Selection of the quantities to be monitored during construction and calculation of the anticipated values based on the most probable conditions;
5. Calculation of the same quantities under the most unfavourable conditions compatible with the available data concerning the ground conditions;
6. Selection in advance of a course of action or modification of the design for every foreseeable significant deviation of the observed conditions from those predicted on the basis of the most probable conditions;
7. Measurement, during construction, of the quantities to be monitored and evaluation of the actual conditions;
8. If necessary, modification of the design to suit the actual conditions.

In the context of EC7, the most probable conditions would be based on the mean values of the parameters while the most unfavourable conceivable conditions (sometimes called the worst creditable conditions) would be assessed directly and would be much more conservative than the characteristic values. The ingredients of the observational method are summarised in **C2.7(1)P** in the form of the following four requirements which must be met:

- Acceptable limits of behaviour shall be established before construction commences;
- It shall be shown that there is an acceptable probability that the actual behaviour will be within these limits;
- Monitoring shall be carried out during construction to check that the actual behaviour is within the acceptable limits; and
- A plan of contingency actions shall be devised which can be adopted should the monitoring reveal behaviour outside these limits.

The advantages of the observational method are that it facilitates designs in situations where the ground conditions are complex or not fully known and it allows optimistic assumptions to be made without increasing the risk above that incurred for designs using the normal method of calculations. Thus this method offers the potential for savings in both time and money and a safer solution on certain types of projects. The risk is reduced in this method by defining the acceptable limits for the behaviour, by preparing the contingency plans before construction commences and then by monitoring the actual behaviour during construction to check that the initial optimistic assumptions have been justified.

The observational method should not be used where rapid deterioration can occur and where there is insufficient time to implement the contingency plans. Also there is no need to follow this method if there is little uncertainty as to how the ground will behave. Examples where the observational method has been used successfully include the design of tunnels, embankments, slopes and propped retaining walls in difficult ground conditions.

2.9   Geotechnical Design Report

An important requirement when designing to EC7 is that the design calculations and the results of the design, as well as the assumptions made and the data on which the design is based, shall be presented in a Geotechnical Design Report (GDR) [C2.8(1)P]. The GDR provides the information contained in what is traditionally termed an interpretative report plus other information and requirements. The level of detail in a GDR will vary depending on the type of design and the complexity of the situation but should normally include the eight items listed in Table 2.7 [C2.8(2)].

Information about the first two items, the site, its surroundings and the ground conditions, is obtained from the geotechnical investigations described in Chapter 3 and is reported in the Ground Investigation Report (GIR). Figure 3.2 shows the particular stages in a construction project when the GIR and the GDR are prepared. Reference should be made in the GDR to the GIR and to other documents which contain more detail about the site and the ground conditions.

EC7 does not state explicitly who should prepare the GDR. However, in C1.4(1)P it is assumed there shall be adequate continuity and communication between the personnel involved in the data collection, design and construction. Hence the GDR should be prepared either by the person who wrote the GIR or by someone else who was involved in the ground investigation.

The need to include a plan of supervision and monitoring, as appropriate, is emphasised in the GDR [C2.8(3)P]. It is also a requirement that the results of the supervision and monitoring checks shall be recorded in an addendum to the GDR. This requirement is a further reflection of the importance in geotechnical designs of continuity in data collection, design and construction.

The final requirement is that an extract of the GDR with the monitoring and maintenance requirements for the completed structure shall be provided to the owner/occupier [C2.8(5)P]. While this is certainly good engineering practice, it has not been common in most countries to provide such requirements and has the potential to cause legal problems. Clients may start looking for cracks in their structures and, if they find any, may, because of the GDR, attribute such cracks to foundation movements, whether this is their cause or not. A corollary to this requirement, which has not been included in EC7, is that designers should retain records of ground investigation data and calculations for a certain period after the

Ref. number	Item
1	A description of the site and surroundings
2	A description of the ground conditions
3	A description of the proposed construction, including the loads and limiting movements
4	Design values of the soil and rock properties
5	Statements on the codes and standards applied
6	Statements on the level of acceptable level of risk
7	Geotechnical design calculations and drawings
8	A list of items to be checked during construction or requiring maintenance or monitoring after construction

Table 2.7: *Items normally included in a Geotechnical Design Report*

end of construction. These should be retained at least until any excess pore water pressures due to loading from the structure have dissipated and any ground movements have effectively ceased.

References

Burland J.B. & Wroth C.P. (1975) Settlement of buildings and associated damage. State of the Art Review, *Proceedings of Conference on Settlement of Structures*, Cambridge, Pentech Press, London.

Duncan J.M., Tan C.K., Barker R.M. & Rojiani K.B. (1989) Load and resistance factor design for highway bridges. *Proc. Symposium on Limit State Design*, Canadian Geotechnical Society, Toronto, pp47-63.

ENV 1991-1 (1994) *Eurocode 1: Basis of Design and Actions on Structures – Part 1: Basis of Design*, CEN Brussels.

ENV 1992-3 (1998) *Eurocode 2: Design of concrete structures – Part 3: Concrete foundations*, CEN, Brussels.

Orr T.L.L. (1994) Probabilistic characterization of Irish till properties, *Risk and Reliability in Ground Engineering*, Thomas Telford, London.

Peck R.B. (1969) Advantages and limitations of the observational method in applied soil mechanics. *Géotechnique*, 19(2) pp171-187.

prENV 1993-5 (1997) *Eurocode 3: Design of steel structures – Part 3: Piling*, CEN, Brussels.

Schneider H.R. (1997) Definition and determination of characteristic soil properties, *Contribution to Discussion Session 2.3, XII International Conference on Soil Mechanics and Geotechnical Engineering*, Hamburg, Balkema.

Simpson B. & Driscoll R. (1998) *Eurocode 7 - A Commentary*, Construction Research Communications Ltd., London.

Student (1908) The probable error of a mean, *Biometrika*, Vol. 6, pp 1-25.

Chapter 3

Geotechnical Investigations and Geotechnical Data

3.1 Introduction

Section 3 of EC7 sets out the requirements for planning geotechnical investigations and obtaining the geotechnical data needed for design. To emphasise the fact that geotechnical investigations to obtain geotechnical data are such an important and integral part of all geotechnical designs, the authors have added the words 'geotechnical investigations' to the title of this chapter compared to the title of Section 3, which is just called 'geotechnical data'. All the reports arising from the different geotechnical investigations at the different stages of a project, and the requirements for these, are thus covered in this chapter.

The three parts of EC7, Part 1: *General Rules* and Parts 2 and 3: *Design Assisted by Laboratory Testing* and *Design Assisted by Field Testing*, should be considered as a whole when assessing the requirements for planning geotechnical investigations and obtaining geotechnical data. Parts 2 and 3 provide the essential requirements for the test equipment and testing procedures, for interpreting and reporting test results, and for obtaining the derived values of soil and rock parameters. Parts 2 and 3 are not testing standards but are part of EC7, which is a standard for geotechnical design. It is intended that Parts 2 and 3 should be used by the designer, not by the persons carrying out the laboratory or field tests.

Section 3 of Part 1: Geotechnical Data is concerned with the design process and sets out the basis for Parts 2 and 3 by giving the general requirements for planning geotechnical investigations, collecting geotechnical data, and presenting and evaluating geotechnical information. The requirements in Section 3 are particularly related to determining a proper description of the ground properties and a reliable assessment of the characteristic values of the ground parameters from the derived parameter values, or the design values, when these are assessed directly. The information obtained from geotechnical investigations, including the

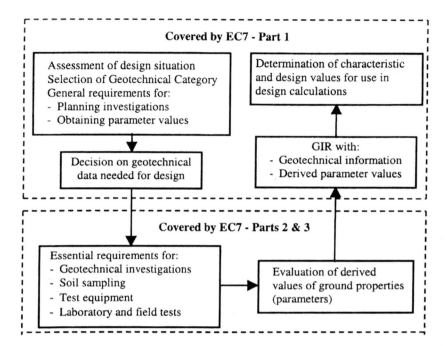

Figure 3.1: *Roles of EC7 - Parts 1, 2 and 3 in obtaining geotechnical data*

geological and groundwater conditions, and the parameter values obtained from the various field and laboratory tests, are presented in a Ground Investigation Report (GIR) as discussed in §2.9. Figure 3.1 shows the link between Part 1 and Parts 2 and 3 with regard to planning geotechnical investigations and obtaining geotechnical data for use in design calculations.

There is some overlap between Parts 1, 2 and 3 of EC7 at present and, as noted in §3.2.3, some inconsistencies. This is understandable considering the development of the EC7. The current intention is to coordinate the contents of different parts, possibly combining Parts 2 and 3 into one document when the review period is complete. The authors would welcome this development.

3.2 Geotechnical Investigations

3.2.1 Scope of Geotechnical Investigations

There is often confusion over the terminology used for the different types of investigations. **C3.2.1(1)P** requires that geotechnical investigations shall provide all data concerning the ground and groundwater conditions at and around the construction site necessary for a proper description of the essential ground properties and a reliable assessment of the ground parameter values to be used in design. Hence, according to this clause, geotechnical investigations are what other codes and engineers often refer to as ground investigations. The terms geotechnical investigations and ground investigations can therefore be taken as being synonymous. Site investigations are generally considered to include other aspects not related to the ground properties, for example environmental conditions or non-technical aspects of the site. However EC7 is not consistent on this point as Section 2 of Part 3 refers to site investigations which appear to be the same as the geotechnical investigations referred to in Part 1.

With regard to the general requirements for geotechnical investigations, **C3.1(1)P** requires that geotechnical information is always carefully collected, recorded and interpreted. This information shall include the geology, morphology, seismicity, hydrology and history of the site. The reference to hydrology is mainly in relation to groundwater hydrology. **C3.1(1)P** also requires that indications of the variability of the ground be taken into account.

According to **C3.1(2)P**, the construction and performance requirements of the proposed structure shall be taken into account when planning geotechnical investigations. **C3.1(2)P** also requires that the scope of the investigations be continually reviewed as new information is obtained during the investigations. As discussed in §2.2.3, this requirement may affect the Geotechnical Category, if these are being used. It is should be noted that the requirements in EC7 apply mainly to GC1 and GC2. Additional investigations of a more specialised nature will often be required for GC3. Thus the investigation requirements in EC7 cannot be taken as being sufficiently comprehensive for complex geotechnical situations such as dams and tunnels.

As **C1.4(1)P** assumes that adequate continuity and communication exist between the personnel involved in data collection, design and construction, it is important that the overall planning and scheduling of both the geotechnical investigations and the geotechnical design be considered at an early stage. To ensure that **C1.4(1)P** is complied with, the parties responsible for the various aspects of a project should be clearly stated, preferably in The Geotechnical Design Report as is recommended in **C2.8(4)** with regard to the parties responsible for making and interpreting the monitoring measurements.

3.2.2 Phases of a Geotechnical Investigation

According to **C3.2.1(2)**, geotechnical investigations should normally be carried out in the following three phases:

- preliminary investigations;
- design investigations;
- control investigations.

Depending on the Geotechnical Category, some of these investigations may overlap; for example, for GC1, the preliminary and design investigations may be combined. Table 3.1 shows the different investigations phases related to the various stages of a project.

Preliminary Investigations

Preliminary investigations are carried out during the planning or feasibility stage

Stage of Project	Investigation Phase	Design Processes	Reporting Requirements
Planning Stage	Preliminary/ Preparatory Investigations	Desk study of ground conditions and walk-over surveys, boreholes, trial pits and probes	No report required by EC7. A <u>feasibility report</u> may be produced at this stage
Design Investigation Stage	Design Investigations	Planning and carrying out laboratory and field test programme in accordance with Parts 2 and 3	<u>Ground Investigation Report (GIR)</u> Presentation and evaluation of all geotechnical information
Design Stage	Further investigations to clarify ground conditions	Geotechnical design to EC7 based on information in GIR	<u>Geotechnical Design Report (GDR)</u> Includes design assumptions, data, design calculations
Construction Stage	Control investigations	Supervision to check actual ground conditions and construction. Monitoring to check the design assumptions, particularly if using observational method	<u>Part of GDR with plan of supervision and monitoring</u> Includes items to be checked or maintained <u>Addendum to GDR</u> with results of supervision and monitoring
Post Construction Stage	None	Monitoring to check structure continues to perform as required. Maintenance	<u>Extract of GDR containing monitoring and maintenance requirements</u> Provided to owner/client.

Table 3.1: *Relationship between project stages, investigation phases and reporting requirements*

of a project. The main purposes of preliminary investigations are, according to **C3.2.2(1)P**, to assess a site's suitability, to compare alternative sites, if relevant, to plan the subsequent design and control investigations and to identify the extent of ground which may have significant influence on the structure's behaviour.

According to **C3.2.2(2)**, preliminary investigations include field reconnaissance (walk-over surveys), desk studies of previous site (i.e. geotechnical) investigations and consideration of construction experience in the vicinity, but no tests are mentioned. In **C2.1(9)P** of Part 3, however, the scope of the preliminary investigations has been expanded from that given in Part 1 to include tests which provide sufficient information to allow decisions be taken on matters such as the positioning of the structure, possible foundation methods, preliminary costs and proposals for the design investigation. This implies that the preliminary investigation, as indicated in Table 3.1, may include boreholes, trial pits and probes to obtain sufficient information for a preliminary design.

Another difference between Part 1 and Parts 2 and 3 is that a fourth investigation phase, termed preparatory investigations, is introduced in Annex A of Part 3 at the start of the project before the preliminary investigations. These preparatory investigations do not involve any tests and so correspond to the preliminary investigations of Part 1, which also do not involve tests. Clearly there is a need for Parts 1 and 3 to be made consistent on this matter.

C2.3(7)P of Part 3 requires preliminary investigations to be performed where the most complicated foundation conditions are expected. C2.3(6)P of Part 3 requires the investigation points to be placed such that an adequate picture is obtained of the variability of the site with regard to the stratification and quality of the ground and groundwater situation.

Design Investigations
According to C3.2.3(1)P, design investigations shall provide all the information required for the design of temporary and permanent works and shall identify any difficulties that may arise during construction. Design investigations normally include borings, in-situ tests and laboratory tests. Where probes or other indirect methods of investigation are used, boreholes should be put down to sample the ground. Part 3 goes into greater detail than Part 1 about what should be included in a design investigation, the methods which should be used and how derived values are determined from the results. For example, C2.4(8)P of Part 3 requires that at least one borehole, with sampling from each layer influencing the behaviour of the structure, shall be made during a design investigation to identify and classify the soils. Part 3 also provides guidance on the quality of sampling, the suitability of the different investigation methods and the interpretation of the results of an investigation.

Control Investigations
Control investigations are investigations carried out to check the actual ground conditions encountered during construction. As these investigations form part of the supervision of construction, the requirements for control investigations are specified in C4.3 of Section 4 on Supervision of Construction, Monitoring and Maintenance and are discussed in §4.2.3.

3.2.3 Determination of Stratigraphy and Extent of Explorations
The purpose of boreholes is to investigate the stratigraphy, to perform field tests and to obtain soil samples for identification and laboratory testing. C3.2.1(2) of Part 1 includes the expected caveat that the extent of geotechnical investigations depends on the Geotechnical Category; i.e. it depends on the ground conditions and type of structure. The following guidelines for planning design investigations for GC2 are presented in C3.2.3(10):

a) For structures which cover a large area, the exploration points, which may be a combination of borings, pits, penetration tests or geophysical soundings, may be on a 20m to 40m grid.

b) For pad or strip foundations, the depth of soundings or borings should normally be between 1 and 3 times the width of the foundation element, with some usually being taken to greater depths to assess settlement conditions and possible groundwater problems. The importance of taking at least some borings to greater depth should be noted.

c) For filled areas and embankments, the minimum depth investigated should include all compressible strata which may contribute up to 10% of the total settlement. The distance between neighbouring exploration points should normally be 100m to 200m.

d) For pile foundations, the depth investigated should normally be at least 5 times the diameter of the pile shaft below the pile toe. In the case of pile groups it is also a recommended that investigations are taken to a depth below the pile toes equal to the smaller side of the rectangle circumscribing the pile group.

These recommendations should be viewed as rough guidelines because comparable experience always needs to be taken into account when planning ground investigations. In cases where the ground conditions are relatively uniform, or where there is extensive local knowledge of the ground and groundwater conditions, these recommendations may prove to be excessive. Part 3 contains no specific recommendations regarding the spacing of exploratory holes but provides more details about their depth. For example **C2.4(10)** recommends that the depth of investigation should extend to all strata that will be affected by the project and **C2.3(3)** recommends that where the overall stability of an area has to be investigated, the ground should be investigated to a depth of at least 5m below potential slip surfaces.

It is interesting to note that, according to **C2.7.2(3)** of Part 3, a rectilinear interpolation between investigation points (i.e. a series of straight lines) may normally be sufficiently accurate when deriving the boundary between layers. This is frequently a contentious point in legal disputes.

According to **C2.4(5)** of Part 3, the investigated area should extend into the neighbouring area around the proposed site for a distance of at least 1.5 times the expected excavation depth or the depth of the soil layer that could generate settlements in the neighbouring area. To satisfy this application rule it would be necessary in many situations to put down boreholes in neighbouring property, which is not always a practical proposition.

3.2.4 Groundwater Pressures

Groundwater pressures have a significant influence on the performance of many geotechnical structures as high pore water pressures cause weakening of soil and increased settlement. Hence the determination of groundwater pressures is most important for design purposes; it is also important for monitoring purposes, to control certain activities during construction, and for risk evaluation. **C3.2.3(11)P** requires that during a design investigation the existing groundwater pressures shall be established as well as the extreme levels of any free water levels that might influence the groundwater pressures. Depending on the design situation, either the upper or lower values of the extreme (ULS) or normal (SLS) groundwater pressures are required for design. Hence, in order to assess correctly the values of the extreme groundwater pressures, it is necessary to observe the fluctuations in the groundwater levels or pressures with time, to assess the hydrogeology of the site, and to identify any artesian pressures. This is particularly important for Case C designs since no partial factors are applied to the water pressure in these. Annex L of Part 3 provides some helpful guidelines for determining design groundwater pressures from groundwater measurements.

To assess an excavation for uplift, **C3.2.3(12)** recommends that the pore water pressures be investigated for a depth below the excavation at least equal to the depth of the excavation below the ground water level, or to greater depths if the material in the upper layers is light. **C2.4(11)** of Part 3 adds the further recommendation that the groundwater pressures be investigated to a depth of at least 3m below the anticipated foundation level in both the high and low points of the area. These recommendations also apply to trench excavations. Details of any dewatering or water abstraction wells in the vicinity also need to be obtained.

Section 14 of Part 3 contains recommendations regarding the most suitable methods for measuring groundwater pressures in different types of ground. However the measurement of negative pore water pressures or the sampling methods for groundwater quality determination are not covered.

3.3   Geotechnical Parameters

3.3.1 Introduction

The process by which the design values of geotechnical parameters are obtained consists of four stages, as shown in Figure 3.2, during which the following four different values are obtained:

1. measured values;
2. derived value of a parameter;
3. characteristic value of a parameter;
4. design value of a parameter.

These four different values are explained below. As indicated by the dashed boxes in Figure 3.2, the requirements that need to be taken into account when evaluating derived values of parameters from the results of laboratory and field tests for use in geotechnical designs to EC7 are covered by Parts 2 and 3 of EC7. The requirements that need to be taken into account when selecting characteristic values and determining design values are covered by Part 1.

Measured Values
A measured value is defined in **C1.4.2.5** of Part 2 as a value measured in a test. Examples of measured values are the water table depth, the blowcounts, N recorded in an SPT test and the stresses and deformations measured in a triaxial test. In some cases, such as the groundwater level, the measured value is the parameter required for design. In other cases, such as SPT tests and triaxial tests, the measured values, such as the blowcounts N or the measured sample stresses and deformations, are only the first stage in obtaining the required parameter value.

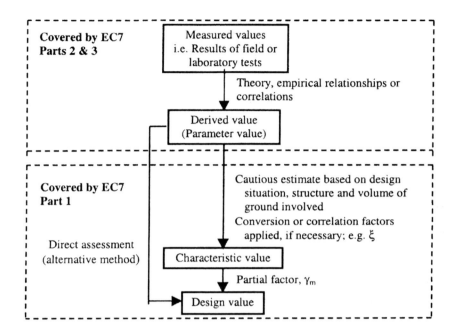

Figure 3.2: *Process for obtaining design values from test results*

Derived Value

A derived value is defined in **C1.4.2.2** of Part 2 as the value of a ground parameter obtained by theory, correlation or empiricism from measured test results. Derived values form the basis for the selection of characteristic values. Examples of derived values are c' and ϕ' values obtained from triaxial compression test measurements using Mohr-Coulomb theory, c_u values derived from field vane test measurements using torque theory, and ϕ' and E_m values obtained from SPT blowcounts using empirical correlations.

It should be noted that the derived value of a parameter is the parameter's value at one particular location in the ground and does not take account of the nature of the structure. It should also be noted that the use of different types of tests can lead to different derived values of the same parameter being obtained at the same location. The example is given in Part 3 of different c_u values being determined by the following three different test methods:

- from CPT test results through correlations with q_c;
- from pressuremeter test results through correlations with p_{LM};
- from laboratory triaxial test results evaluated theoretically.

Each of the above methods will give different derived values for c_u at a particular point. It is important, therefore, when reporting derived values, to clarify how the derived values have been obtained and state what assumptions have been made.

According to **C3.3.1(3)P**, the items that need to be considered in order to obtain reliable values of geotechnical parameters from a particular test are:

- stress level, mode of deformation, etc.;
- relevant published material for the test in appropriate ground conditions in order to interpret the test results;
- testing schedules to include a sufficient number of tests to provide data for the derivation and variation of the relevant parameters;
- comparison of each parameter value with published data, local experience and large scale field trials, if available;
- correlations between the results of more than one type of field test, whenever available.

The concept of derived values has given rise to some debate. In some cases the distinction between measured test results and derived values is clear, as in the case of ϕ' values derived by correlation from the N values measured in SPT tests. In other cases, however, the distinction is not so clear. For example in a CPT test the strains in the instrument are measured during the test and then, without the operator being involved, the cone resistance q_c is calculated from the strains. Thus the q_c value is in fact a derived value, but it could also be argued that it is a test result or measured value. For this reason it has been proposed by some that the term 'parameter value' should be used for a value obtained from test results.

Characteristic Value
As explained in §2.5.3, the characteristic value of a geotechnical parameter is defined as a cautious estimate of the value affecting the occurrence of the limit state. The characteristic value, therefore must be selected taking account of the actual design situation. **C3.3.1(1)P** emphasises this by requiring that field and laboratory test results be interpreted in a manner appropriate to the limit state being considered. The clauses of Section 3 draw attention to those aspects of field and laboratory tests that need to be considered when selecting characteristic values.

In the case of some field and laboratory tests, a conversion (or correlation) factor is applied, if necessary, to convert the test results into values which can be assumed to represent the behaviour of the soil or rock in the ground at the actual limit state or to take account of uncertainties in any correlation used to deduce derived values from test results [**C2.4.2(3)**]. An example of a conversion factor is Bjerrum's (1972) correction factor which is used to convert the c_u value measured in a field vane test into a value that represents the actual shear strength of the ground deforming under plane strain conditions.

Correlation factors (ξ) are used to obtain characteristic pile resistances from the results of pile load tests as shown in §7.5.3 and §7.6.2. These factors take account of uncertainties in the load test results and the variability of the ground conditions.

Design Value
As explained in §2.5.2, the design value is the value of the parameter used in design calculations. It is determined, in accordance with **C2.4.3(1)P**, either from the characteristic value by applying a partial factor or else directly by assessment from the derived value as shown in Figure 3.2.

Section of Part 3	Field test	Derived parameters
3	Cone penetration and piezocone tests – CPT & CPTU	c_u, ϕ', E_m, E_{oed}
4	Pressuremeter test - PMT	ϕ', E_M, p_{LM}
5	Standard Penetration Test - SPT	N, I_d, ϕ'
6	Dynamic Probing Test - DPT	N_{10}, N_{20}, r_d, q_d
7	Weight Sounding Test - WST	c_u, ϕ', E_m,
8	Field Vane Test - FVT	c_u
9	Flat Dilatometer Test – DMT	p_0, I_{DMT}, K_{DMT}, E_{DMT}, c_u, ϕ', E_{oed}
10	Rock Dilatometer Test – RDT	E_d, E_{d1}
11	Plate Loading Test - PLT	c_u, E_{PLT}, k_s

Table 3.2: *Field tests and derived parameters*

3.3.2 Field Tests

The most important requirements to be considered when planning field tests are presented in Part 3. These include the requirements for the equipment, for carrying out the tests, for reporting the results, for interpreting the results and for obtaining the derived values of geotechnical parameters. Table 3.2 provides a list of the in situ tests covered by the different sections of Part 3, and the parameters obtained from these, for which essential requirements are given. It should be noted that Part 3 is not a testing standard and so does not provide detailed specifications for the field tests. For these reference should be made to national or international field testing standards such as, for example, BS 1377 (1990).

3.3.3 Soil Sampling

Some guidance on the spacing of sampling points in a borehole is given in Part 3 which requires that, for identification and classification of soils, samples shall be obtained from every separate layer influencing the behaviour of the structure [C2.4(8)P] and, for organic soils, samples should be taken at least every 1m in one boring [C2.4(9)].

The requirements for sampling soil and rock, including the requirements for samplers, sampling techniques and sampling procedures and the requirements for the handling and storage of samples and the reporting of these procedures, are given in Sections 12 and 13 of Part 3. There is a wide variety of sampling methods and sampling procedures used throughout Europe. In order to differentiate between the different sampling methods, they are classified in C12.2.1 of Part 3 into following three categories:

- *Category A*   Those sampling methods where no, or only slight, disturbance of the soil structure occurs during sampling;
- *Category B*   Those sampling methods which retain the constituents of the in situ soil, its natural water content and the general arrangement of the layers, but the structure may be disturbed;

- *Category C*   Those sampling methods where the soil's structure is totally changed and the water content may not represent the natural water content of the sampled soil layer.

The sampling method category depends not only on the details of the sampler used, for example on its area ratio and inside clearance ratio as defined in **C12.2.2(1)P** of Part 3, but also on the sampling procedure. Guidance on the sampling methods with respect to the sampling category achievable for the different soil types is given in Annex J of Part 3.

C2.3(1) of Part 2 describes five quality classes of soil samples in relation to laboratory testing which depend on the characteristics that remain unchanged on sampling. These should not be confused with the three sampling method categories described above. The following are the five quality classes of samples:

- *Quality Class 1*   Sample where no change in characteristics of practical significance has occurred (undisturbed);
- *Quality Class 2*   Sample where some disturbance has occurred. The disturbance could, for example, influence the results of shear strength and compressibility tests (slightly disturbed);
- *Quality Class 3*   Sample which has retained all the original constituents of the soil in situ, including water content;
- *Quality Class 4*   Sample which has retained all the original constituents of the soil in situ but where the water content may not represent the in situ value;
- *Quality Class 5*   Sample where the structure, water content and constituents of the soil have been changed during sampling so that it is not suitable for representative testing and only suitable for determining the sequence of the strata (severely disturbed).

It should be noted that the quality class of the sample for laboratory tests obtained by a certain sampling method depends very strongly on the details of the sampler and the quality of the sampling procedure.

When planning geotechnical investigations, it is important to realise that a particular sample quality required for testing can only be obtained using certain sampling method categories. The relationship between sample quality classes and sampling method categories, given in Table 12.1 of Part 3, is as follows:

- Sample Quality Classes 1 and 2 are only obtained from Category A methods;
- Sample Quality Classes 3 and 4 are only obtained from Category B methods or higher;
- Sample Quality Class 5 is obtained from Category C methods or higher.

The minimum sample quality classes that the person planning a testing programme should specify for the test samples in order to obtain particular soil properties and soil parameters are shown in Table 3.3. This table shows, for example, that a Quality Class 5 sample is only needed to determine the sequence of layers, i.e. this property can be obtained from all sample qualities. However a Quality Class 1 sample is recommended to obtain the compressibility and shear strength. Samples with lower quality classes than those shown in Table 3.3 may be used to determine some parameter values, but then the effects of sample disturbance need to be considered in the interpretation of the results.

Soil properties	Soil parameters	Minimum Sample Quality Class
Sequence of layers	-	5 (severely disturbed)
Particle size, Atterberg limits, particle density, organic content, boundaries of strata (broad)	w_L, w_P, G_s,	4
Water content	w	3
Density, density index, porosity, permeability, boundaries of strata (fine)	ρ, I_D, n, k,	2
Compressibility, shear strength	C_c, C_s, m_v, E_{oed}, c_v, c', ϕ', c_u	1 (undisturbed)

Table 3.3: *Sample Quality Classes recommended to obtain soil properties and parameters*

3.3.4 Laboratory Tests

The most important requirements to be considered when planning laboratory tests to obtain ground parameter values are presented in Part 2. These requirements include those relating to the testing equipment, the way the specimens for testing are prepared, the testing procedure and the reporting of the results. Guidance is also provided on the planning of a laboratory testing programme, on evaluating and reporting test results, and on obtaining derived values of geotechnical parameters from different types of tests. The types of tests covered in the different sections of Part 2 are listed in Table 3.4.

It should be noted that Part 2, like Part 3, is not a testing standard and so does not provide detailed procedures for carrying out laboratory tests. It does, however, in Annex B, give a list of appropriate national and international testing standards. Included in this list is the ISSMGE (1999) document, referred to in §1.3.1, that provides recommendations for the most common geotechnical laboratory tests. Other annexes provide recommendations and guidelines on developing testing programmes, on the mass of soil/rock required for each test, on the calibration regimes and on other aspects of laboratory testing. While these recommendations are generally very practical and helpful, some may not be generally applicable. For example, according to **A9.2.1(2)** of Part 2, a shear box where the two halves of the shear box move exactly in parallel should be used to prevent tilting. This facility is not available in many of the direct shear boxes currently on the market. If the two plates are not parallel, it is stated that the friction angle obtained can be as much as 4° higher in clay soils and as much as 6° lower in sands.

When planning a laboratory testing programme, **C2.2(1)P** of Part 2 requires that the soil type, stratigraphy, type of structure and geotechnical aspects of the project be taken into account. All soil samples should be inspected visually prior to any laboratory test being performed to establish a preliminary soil profile [**C2.2(3)** of Part 2]. This visual inspection should be supported by simple manual tests to classify and identify the soil. The results of these tests shall be assessed as they

Section of EC7 Part 2	Laboratory tests
5	Tests for classification, identification and description of soils
6	Chemical testing of soils and groundwater
7	Compressibility testing of soils
8	Strength index testing of soils
9	Strength testing of soils
10	Compaction testing of soils
11	Permeability testing of soils
13	Classification testing of rock material
14	Swelling testing of rock material
15	Strength testing of rock material

Table 3.4: *Sections of Part 2 covering laboratory tests for soils and rock*

become available and the testing programme adjusted where necessary [**C2.2(8)P**].

The following terms, which are commonly used in relation to the laboratory testing of soil, are defined in **C1.4.2(2)P** of Part 2:

- Specimen: a part of a soil or rock sample used for a laboratory test;
- Element test: a test on an element (specimen) of soil to determine a property.

Other terms, such as remoulded, re-compacted, re-constituted and re-consolidated specimens, are also defined in **C1.4.2(2)P** of Part 2. An important definition introduced in **C1.4.2(15)P** of Part 2 is the term 'strength index test' which covers rudimentary tests that yield an indication of the shear strength. It is stressed that the results of such tests are subject to considerable uncertainty. The laboratory vane, pocket penetrometer, fall cone, unconfined compression and unconsolidated undrained compression test all fall within this category. Inclusion of the unconsolidated undrained compression test may be considered by some to be a downgrading of the c_u value obtained from this test, which is frequently used in geotechnical design. It is, however, an acknowledgement of the fact that loss of suction occurs on sampling in some soils and this can significantly affect the c_u value measured in an undrained compression test.

3.3.5 Quality Control

Quality assurance and quality control are only specifically mentioned in Part 2 in relation to laboratory testing. Both Parts 2 and 3, however, give detailed checklists and calibration periods which shall be adopted for laboratory and field testing equipment. **C2.6(1)P** of Part 2 requires that an appropriate quality assurance system shall be in place in the laboratory and engineering office, and that quality control shall be exercised competently in all phases of laboratory testing and in the interpretation of the test results.

A very important requirement in **C2.6(2)P** of Part 3 regarding the quality and reliability of geotechnical data should be noted. This is the requirement that all site (i.e. geotechnical) investigations shall be reported in such a way that the results can be checked and re-evaluated by persons other than the author.

3.3.6 Evaluation of Test Results and Geotechnical Parameters

As mentioned in §3.3.1, Parts 2 & 3 give the requirements for evaluating derived values from laboratory and field tests. Parts 2 and 3 also give guidance on the presentation and evaluation of test results; for example **C2.5(2)P** of Part 2 requires that the evaluation of laboratory test results takes into account available comparisons of the results with existing experience, the results of other types of laboratory tests, and correlations based on index properties of the soil. In the case of drained triaxial compression and direct shear tests, **C9.3(4)P** of Part 2 requires that the stress range over which the strength parameters in terms of effective stresses have been determined be given. **C9.3(3)** notes that linear interpolation of test results using Mohr's circles can give erroneous values of the soil strength as the Mohr failure envelope is not a straight line, especially at low normal stresses.

In relation to the evaluation of parameter values, **C1.4.2.1** of Part 2 expands the definition of comparable experience in **C1.5.2(1)P** of Part 1 to include three classes of experience:

*   none: if no reliable results are available;
*   medium: if data for similar soils are available; and
*   extensive: if statistical evaluations and/or published correlations exist, or if test results for the same soil on a nearby location exist.

C2.7 of Part 3 provides some useful guidance on the evaluation of information obtained from geotechnical investigations and requires that the evaluation be based on all the information available from the different phases of the investigation. The evaluation of geotechnical information should include the following:

*   tabular and graphical presentation of field and laboratory test results;
*   determination of the depth and seasonal fluctuation of the ground water level;
*   subsurface profiles showing the boundaries of the different soil and rock strata;
*   the range of values of the geotechnical data for each stratum.

C3.3 of Part 1 provides the requirements for evaluating geotechnical parameters. What is intended by this is the evaluation of characteristic parameter values from derived values or the evaluation of design values if they are being assessed directly rather than through the application of partial factors. This evaluation involves taking account of factors such as the limit state being considered, the volume of soil involved in the limit state, the stress level in situ compared to that in the test, and the mode of deformation involved. The distinction between evaluating derived values from test results and evaluating characteristic values from derived values is not clear in the present texts of Parts 1, 2 and 3, but it is anticipated that revision of these texts will clarify this aspect.

The sub-clauses of **C3.3** are very valuable in that they highlight the aspects that need to be considered when evaluating geotechnical parameters from test results. For example **C3.3.6(1)P** lists the factors, such as the differences between the stress state in situ and in the test, the effect of sample disturbance, anisotropy, fissures and rate effects, that shall be considered when evaluating the undrained shear strength of a fine grained soil. **C3.3.7(3)P** requires that, when assessing c' and ϕ' values from the results of undrained triaxial tests with pore pressure measurements, attention be paid to the fact that the samples are fully saturated. Another factor to be considered when assessing ϕ' values is the fact that soils generally exhibit slightly higher ϕ' values when tested in plain strain than under triaxial conditions [C3.3.7(4)]. Hence, when designing retaining walls for example, the ϕ' value derived from triaxial tests may be increased by 2° to 3° to allow for plane strain conditions. Such an increase would not be appropriate in design situations where plane strain conditions do not prevail. When a parameter value has been derived using a correlation, it is important in assessing the design value to take into account whether the correlation relates to the mean or a conservative value of the parameter.

3.4   Ground Investigation Report

The results of preliminary and design investigations shall be compiled in a Ground Investigation Report (GIR) [C3.4(1)P]. The GIR should normally consist of two parts, one part which presents all the available geotechnical information and another part which provides the derived parameter values and an evaluation of the information. The GIR forms the basis of the Geotechnical Design Report (GDR) described in §2.9, which also contains the reporting requirements for the control investigations. The relationship between the different investigations and the reports required is shown in Table 3.1.

The factual report part of the GIR should contain a factual account of all the field and laboratory work and details of the test methods used. A list of items that should be included is provided in C3.4.1(2) and includes the project details project, the names of the companies involved in the investigation, local experience in the area, borehole logs, presentation of test results, plus many other items.

The second part of the GIR is the evaluation of the geotechnical information and is covered by C3.4.2. This evaluation of the geotechnical information is different from the evaluation of the test results and derived parameter values discussed in §3.3.6 above. Evaluation of the geotechnical information involves a review of the field and laboratory work and an evaluation of the quality, relevance, accuracy and sufficiency of the geotechnical information. Any inadequacies in the information are to be highlighted and commented upon. Detailed proposals for further investigations are to be given, if deemed necessary.

References

Bjerrum, L. (1972) Embankments on soft ground, *Proceedings ASCE Speciality Conference on Performance of Earth and Earth-Supported Structures*, Perdue University, Layfayette, Indiana, pp 1-54

BS 1377 (1990) *Methods of test for soils for civil engineering purposes,* British Standards Institution, London

ISSMGE (1999) *Recommendations of the ISSMGE for geotechnical laboratory testing*, DIN-Beuth Verlag, Berlin.

Chapter 4

Supervision of Construction, Monitoring and Maintenance

4.1 Introduction

It may appear strange to many engineers that a code about geotechnical design should have a section with requirements for the supervision of construction, for monitoring the performance of a structure and for the maintenance of a structure. The reasons for these requirements are the uncertainties in geotechnical designs concerning the actual ground conditions and the objective of the Eurocodes, which is to achieve safe designs. Hence, having designed a structure, for example a foundation, using assumptions about the ground conditions based on a limited number of field and laboratory tests, it is necessary to inspect the actual ground conditions and observe the performance of the structure to check if the assumptions are valid. Section 4 of EC7 covers this aspect of the design process. **C4.1(1)P** requires that the following activities shall be carried out, as appropriate, to ensure the safety and quality of a structure:

- the construction processes and workmanship shall be supervised;
- the structure's performance shall be monitored during and after construction;
- the structure shall be adequately maintained.

EC7 is primarily concerned with the designer's responsibility for specifying the requirements for construction supervision, monitoring and maintenance. The requirements for workmanship are covered by the execution codes listed in §1.3.2.

How and where supervision, monitoring and maintenance fit into the geotechnical design process is shown in Table 3.1. The supervision, monitoring and maintenance requirements in EC7 are qualified by the words "as appropriate" because the amount of supervision, monitoring and maintenance required in any particular case will depend on the nature of the structure and on the ground conditions. In the case of GC1, visual inspection of the ground, rough quality controls and a qualitative assessment of the structure's performance may be all that is necessary. Monitoring and maintenance may not be required at all. An example of this would be the foundations for a small structure on good ground. For GC2, measurements of the properties of the ground encountered or the behaviour of structures may often be required. Tests to check the performance and quality of piles and tests to check the density of backfill behind retaining structures are examples of monitoring that may be required for GC2. For GC3, for example an earth embankment on soft ground, additional measurements of the ground and structure may be required during each significant stage of construction. Also the post-construction performance of the structure may need to be monitored and parts of the structure, for example the slopes of an embankment, may need to be maintained during the life of the structure.

Who is to do the Supervision, Monitoring and Maintenance?
Who should actually do the supervision, monitoring or maintenance is not stated in EC7. **C1.4(1)P** just notes it is assumed that adequate continuity and communication exist between the personnel involved in the design and the construction and that the structure will be adequately maintained. Ideally the person who prepares the design should also supervise the construction. However in practice this will not always be the case. Therefore there is a need to ensure that good communication exists between those involved in the design and those involved in the construction and the maintenance of a structure. The way this communication is established when designing to EC7 is by means of the GDR.

C2.8(3)P requires that the GDR shall include a plan of supervision and monitoring with items to be checked during construction or which require maintenance after construction. **C2.8(5)P** requires that an extract of the GDR with the monitoring and maintenance requirements for the completed structure be provided to the owner/client. Items which require checking during construction or monitoring after construction are discussed below in §4.2 and §4.3 respectively.

4.2   Supervision

4.2.1 Supervision and Checking
Supervision in EC7 means checking the design and the construction. Hence the purposes of supervision as part of the design process are to:
- check the validity of the design assumptions by comparing the actual conditions with those assumed in the design;
- identify the differences between the actual ground conditions and those assumed in the design so that the design principles can be assessed in the light of the actual conditions;
- check that the construction is carried out in accordance with the design.

The plan of supervision included in the GDR should clearly identify the items that need to be checked during construction and should state the acceptable limits for the results obtained from the supervision.

Annex A.1 of EC7 provides a list of the more important items to be considered

Type	Items to be checked	Checked
1a	Ground conditions	
1b	Location and general layout of structure	
2a	Groundwater flow and pore pressure regime	
2b	Effectiveness of measures to control pore pressures and seepage	
2c	Efficiency and effective operation of dewatering system	
2d	Control of dewatering to avoid disturbing neighbouring structures	
2e	Chemical composition of groundwater and corrosion potential	
3a	Structural settlements and stability of excavation walls and bases	
3b	Temporary support systems	
3c	Effects on nearby buildings and utilities	
3d	Soil pressures on retaining structures	
3e	Pore pressure variations resulting from excavation or loading	
4	Safety of workers with regard to geotechnical limit states	

Table 4.1: *More important items to be checked when supervising construction*

when supervising construction and these are summarised in Table 4.1. The items in Table 4.1 have been divided, by the authors, into four different types: Type 1 is concerned with the ground conditions, Type 2 with the groundwater, Type 3 with movements and pressures and Type 4 with the safety of workers. Most of the items in Table 4.1 are concerned with checking the ground and the groundwater conditions. It should be noted, however, that Type 4 is concerned with workers' safety. Generally workers' safety is not covered by the Eurocodes but is covered by national health and safety regulations. Workers' safety has been included in EC7 because of the responsibility of the person checking construction to ensure that workers will not be endangered during construction by the occurrence of geotechnical limit states, for example by the collapse of a trench.

4.2.2   Inspections and Keeping Records

According to **C4.2.2(1)P** the construction work shall be inspected visually on a continuous basis and the results of these inspections shall be recorded. What is meant by a 'continuous basis' is that inspection should be on-going during the construction process and should not just be infrequent isolated inspections as this could result in important features of the ground conditions not being identified. The requirement that records shall be kept of all observations and measurements is an important feature of EC7. These records should be kept in a systematic manner, so that they can be readily interpreted, and they should be made available to the designer. **C4.2.2(3)P** requires that the inspection shall result in records of the following, as appropriate, being maintained:

- significant ground and groundwater features;
- sequence of works;
- quality of materials;

- deviations from design;
- as-built drawings;
- results of measurements and their interpretations;
- observations on the environmental conditions;
- unforeseen events.

4.2.3   Checking Ground Conditions – Control Investigations

Checks of the ground and groundwater conditions during construction are the third type of geotechnical investigations and are referred to in **C3.2.1(2)** as control investigations. Control investigations involve:

- checking the assumptions in the design regarding the descriptions and geotechnical properties of the soils and rocks in or on which the structure is founded and, if necessary, carrying out further investigations [**4.3.1(1)P**];
- checking, as appropriate, the groundwater levels, pore pressures and groundwater chemistry encountered during construction and comparing them with those assumed in the design [**4.3.2(1)P**]. The effects of constructions operation, including dewatering, grouting and tunnelling, on the ground water regime needs to be checked [**C4.3.2(3)**].

For GC1 the control investigation will normally involve just a visual inspection of the ground exposed during excavations to check the descriptions of the ground conditions [**4.2.2(2)**]. An example is the inspection of the ground upon which a foundation is to be placed. This inspection should check that the actual bearing stratum is the ground assumed in the design, that it has adequate strength and has not been disturbed or softened as a result of construction activity or the weather.

For GC2, the control investigation should include checks on the properties of the soil or rock in or on which the structure is founded. Hence for GC2 structures, additional investigations may be required and samples may need to be taken for laboratory testing as part of the control investigation. Indirect evidence of the geotechnical properties of the ground (e.g. pile driving records) should be recorded and used to assist in the interpretation of the ground conditions [**C4.3.1(2)**].

4.2.4 Checking Construction

As well as checking the ground and groundwater conditions, **C4.4(1)P** requires that the construction operations are checked to ensure that they comply with the construction methods assumed in the design and stated in the GDR. The application rule to this code requirement, **C4.4(2)** notes that for GC1 the construction schedule and sequence of construction is not normally included in the GDR, while for GC2 and GC3, the GDR may give the sequence of operations envisaged in the design or may state that that the sequence of construction is to be decided by the contractor. Traditionally in many countries, to avoid legal problems, the methods of construction are not specified by the designer but are left to the contractor. However the introduction of the **C4.4(1)P** requirement should result in designers having to consider more carefully the likely construction methods and having to specify in the GDR the methods assumed in the design.

An example of a check on construction is the checking of fill placement and compaction discussed in §5.2.6 and the requirements for which are given in **C5.3.4**. Another example of a construction check is the requirement that the GDR shall specify checks to verify that the unit weight of the backfill behind a retaining wall is no worse than the value assumed in the design [**C8.3.1.1(1)P**].

4.2.5 Reporting Deviations from the Design Assumptions

The same requirement is repeated in **C4.3.1(3)P**, **C4.3.2(4)P** and **C4.4(3)P** that deviations from the ground type and properties, groundwater conditions and construction methods assumed in the design shall be reported without delay to the person responsible for the project. This is an important principle but its interpretation with regard to what exactly is meant by "without delay" may be a source of legal problems in the future. What is required is that clear lines of communication are established to ensure that, should any deviations from the design assumptions be discovered which could cause problems, these are reported as soon as possible to the person responsible.

4.3 Monitoring

Monitoring in EC7 involves observing and measuring the performance of a structure. Two aims of monitoring are mentioned in **C4.5(1)P**. The first aim is to check the validity of the assumptions made in the design or as an aid to the design process during construction. When monitoring is an aid to the design process it is part of the observational method and is discussed in §2.8. The second aim of monitoring is, post-construction, to check that the completed structure will continue to perform as required after construction. The length of post-construction monitoring may be altered as a result of observations during construction [**C4.5(4)**].

When monitoring is required to ensure that a structure will continue to perform as required after construction, the required inspections and measurements shall be specified in the extract of the GDR provided to the owner/client [**C4.5(2)P**]. According to **C4.5(4)**, the measurements that may be required when monitoring a structure include:

- deformations of the ground affected by the structure;
- values of actions;
- values of contact pressure between the ground and the structure;
- pore water pressures and their variations with time;
- stresses and deformations in structural members.

Annex A.2 provides a list of the more important items to be considered when monitoring the performance of a completed structure and these are summarised in Table 4.2. The type of monitoring chosen to evaluate the performance of a structure will depend on the geotechnical category. According to **C4.5(6)** a simple visual inspection may only be required for GC1, while for GC2 monitoring may involve measurements taken at selected points in the structure, and for GC3 monitoring should normally be based on measurements of displacements and

Items to be considered	Checked
Settlement at established time intervals of buildings and other structures	
Lateral displacement and distortions, especially those related to fills, stockpiles and soil supported structures	
Piezometric levels	
Deflection or displacement of retaining structures	
Flow measurement from drains	
Temperature monitoring and ground movements in the case of high or low temperature structures	
Watertightness	

Table 4.2: *More important items to be considered when monitoring performance*

analyses which take account of the construction sequence.

C4.5(5)P requires that all measurements shall be evaluated and interpreted; there is no value in taking measurements unless these are evaluated and their implications assessed. The interpretation of the inspections and measurements shall be carried out by an appropriately qualified person, as assumed in **C1.4(1)P**, and, in accordance with **C2.8(4)**, the person responsible should be stated in the GDR. This person shall take appropriate action should the structure not be performing as intended in the design. Also, although only required by **C2.7(1)P** when monitoring is part of the observational method, contingency plans of action should be prepared in case the monitoring of the completed structure reveals that it is not behaving as anticipated in the design.

4.4 Maintenance

In the case of many geotechnical structures, such as spread and pile foundations, maintenance in the long term, after construction, is not normally required. However there are situations, for example in the case of slopes and retaining structures, particularly when groundwater is involved, where some maintenance may be required to ensure both the safety and the serviceability of the structure. In these situations it is necessary to specify in the GDR extract provided to the owner/client [**C4.6(1)P**] the maintenance requirements. These requirements should include the parts of the structure to be inspected and the frequency of the inspections [**C4.6(2)**]. Three examples of maintenance requirements are:

1. Inspections of the drains behind a retaining wall, every six months say, to prevent blockage of the drains causing build up of pore water pressure resulting in unacceptable movements or instability of the wall;
2. Inspections of the slopes of a cutting, once a year say, to ensure that erosion due to rain or groundwater does not remove soil thereby increasing the inclination of the slope and reducing the stability of the cutting;
3. Inspections of the slopes of a river embankment, every two years say, to ensure that trees which could endanger the stability do not become established.

Chapter 5

Fill, Dewatering, Ground Improvement and Reinforcement

5.1 Introduction

The requirements for the design of fill, dewatering, ground improvement and (ground) reinforcement to EC7 are combined in one section, Section 5. This is because each of these processes is concerned with improving the properties of the ground, while the previous section, Section 4, is concerned with determining the properties of existing ground. Fill, dewatered and improved ground all have to satisfy the same fundamental requirements as existing ground, which are that they shall be capable of sustaining the actions caused by loads, seeping water, vibrations, temperature, rain etc. [C5.2(1)P]. Section 5 provides the general requirements for the design of fill, dewatering and ground improvement, giving few specific requirements except for some aspects of fill. This is reflected in the text of this chapter, which is mainly concerned with fill. The design procedures for geotechnical structures which may involve the use of fill or ground improvement processes are covered in the sections of EC7 dealing with the design of spread foundations, piles, retaining structures, and embankments and slopes.

5.2 Fill

5.2.1 Design of Fill

The design of fill beneath foundations and floor slabs, as well as backfill to excavations and retaining structures, is covered by the provisions in Section 5. The design of fill for general landfill, including hydraulic fill, landscape mounds, spoil heaps and fill for embankments for dams (dykes) and transport networks are also

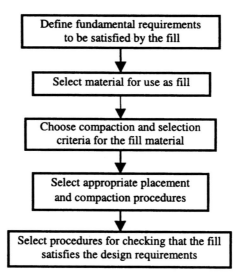

Figure 5.1: *Sequence of steps in the design of fill*

covered [**C5.1(2)**]. The terminology adopted by the authors to describe the different types of fill is that a fill designed to meet specific requirements is termed an 'engineered fill', while that placed in an undesigned and uncontrolled manner is referred to as 'made ground'. When an engineered fill is designed to support a structure, for example a foundation, it is termed a 'structural fill', as in **C5.3.4(2)**.

The sequence of steps involved in the design of fill to EC7 is shown in Figure 5.1 and includes:

- determining the design requirements for the fill;
- selecting material for fill;
- choosing compaction criteria to ensure the fill satisfies the design requirements and selection criteria to ensure the selected material can be compacted to meet the compaction criteria;
- selecting appropriate placement and compaction procedures; and
- selecting procedures for checking that the fill satisfies the design requirements.

Some of the different aspects involved in these steps are interlinked; for example the selection of suitable material for use as fill material depends not only on the type of material and the design requirements, but also on the placement and compaction procedures to be used. Perhaps the most important aspects in the design of fill are the criteria for selecting the material to be used and for the compaction to ensure that the fill satisfies the design requirements.

Checking the placement and compaction of fill during the construction stage to ensure that it complies with the compaction criteria and so satisfies the design requirements is an example of the control investigations mentioned in §4.2.3.

5.2.2 Fundamental Requirements for Fill

The first step in the design process for fill is determining the fundamental requirements to be satisfied by the fill which, as noted above, are that it shall be capable of sustaining the actions caused by loads, seeping water, vibrations, temperature, rain etc. To satisfy these requirements the strength, stiffness and durability of the fill shall be sufficient to support the design loads and ensure that the movements during the structure's design life due to these loads, moisture changes or particle degradation are within the design limits. If the fill is required to act as a water barrier, there may be a requirement that there is an upper limit to its permeability, or if it is to be fill for the abutment of an integral bridge, an upper limit may be placed on the angle of shearing resistance of the compacted material.

Specific requirements are given in **C5.3.4** regarding an appropriate dry density for suitable fill on which foundations are to be built. However specific requirements for fill for other civil engineering works, for example road embankments constructed of cohesive fill, are not provided. This is not surprising as the Eurocodes were originally intended to provide a group of harmonised standards for the structural design of buildings.

Fill is required to have good transportation, handling and placement properties as well as adequate engineering properties after compaction. Another requirement, in the case of fill for road construction for example, is the trafficability of the fill. This is the requirement that construction equipment can traffic on the fill and that movement under the equipment required to lay subsequent pavement layers is within the acceptable tolerances.

The design of fills in many cases involves making the maximum use of available resources. If the fundamental requirements cannot be satisfied with the available fill, ideal foundation conditions may have to be compromised to make projects economically feasible. For example, in marine land reclamation it is frequently not economically viable to design a fill on which structures can be placed using pad foundations. In such a case, structures can be founded on piles or on locally improved ground rather than on the fill itself, while the settlements of roads on the fill can sometimes be tolerated. The selection of the fundamental requirements is therefore a critical part of the overall design process.

5.2.3 Fill Selection

Having defined the fundamental requirements to be satisfied by a fill, it is then necessary to determine the criteria for selecting material that is suitable for use as fill for the particular design situation. **C5.3.2(3)P** lists the aspects which shall be considered when selecting material for use as fill. These include the grading of the material, its compactibility, organic content, susceptibility to volume change, resistance to weathering, resistance to frost and other considerations. Admixtures, crushing, protection or the use of drainage layers to improve the fill may be considered if the material is not suitable in its natural state [**C5.3.2(4)**]. Care is required when selecting fill material with potentially aggressive or polluting chemicals [**C5.3.2(5)P**] and material shall not be used if it contains significant amounts of snow, ice or organic material [C5.3.2(8)P].

Type of fill	Cohesive fill	Granular fill	Rock fill
Particle sizes	Fines content > 15%	Fines content < 15%	65% > 25mm

Table 5.1: *Proposed definitions for three types of fill based on particle size*

Types of fill material
The recommendations in C5.3.2(2) regarding the types of material that are suitable for use as fill mainly relate to structural fill. According to this clause, suitable fill materials include most graded natural granular material and certain waste products, such as selected colliery waste and pulverised fly ash (PFA). It also mentions that some manufactured materials, such as light aggregate, can be used in some circumstances and that some cohesive materials may be suitable but require particular care. This list does not consider fills for use as impermeable barriers or fills for road embankments, where the use of cohesive material would be the norm rather than the exception.

Grading of suitable fill material
While the general principles of C5.3.2 are clear regarding the type of material that is suitable as fill, the grading of suitable fill material is not clear as the terms 'graded', 'granular' or cohesive' are not defined. Also the use of rock fill material is not mentioned in this clause, although it is mentioned in C5.3.4(2) in relation to the checking of fill. To remedy this deficiency, the authors propose the following definitions and classifications for the three types of fill which are summarised in Table 5.1.

Cohesive fill
The authors propose that a cohesive fill is defined as a material with a fines content (% of particles passing the 0.06mm sieve) greater than or equal to 15%. This is the upper limit for the fines content of a material which can readily drain under gravity. As a cohesive material does not drain readily, its water content cannot be reduced during construction in a reasonable time period. The 15% fines content is also considered to be a reasonable boundary between those soils which are likely to undergo significant volume changes due to variations in suction pressure during construction and those which are not. A cohesive material is only suitable for use as fill if its water content, when placed, is such that it can be compacted to satisfy the fundamental requirements.

Granular fill
The definition proposed for granular fill is a material that has less than 15% fines; however its grading shall also be satisfactory [C5.3.2(3)P]. This means that the grading shall be such that it can be compacted and, when compacted, it is internally stable and no migration of fines occurs. The latter point is particularly important where water is likely to flow through the fill.

Rock fill
Rock fill is material in which the fines do not fill the voids and hence the strength

and durability of the particles have a significant effect on the long-term performance of the fill. Leps (1973) has suggested that the boundary between a rock fill and what he described as an 'earth fill' is when about 30% to 35% of the material by weight passes the 25mm sieve. The authors consider that this may be taken as a useful guide for defining rock fill.

Durability of fill material
Although not mentioned in Section 5, the durability of fill needs to be considered [C2.3(1)P], particularly for rock fill placed in marine conditions. There are two aspects of durability: 1) mechanical deterioration or wear, and 2) physio-chemical deterioration. The Aggregate Abrasion Value Test (BS 812, 1990) is an example of a test for mechanical deterioration. The most common tests for physio-chemical deterioration are the Slake Durability Test (ISRM, 1981) and the Magnesium Sulphate Soundness Test (BS 6349, 1984). These are usually accompanied by a petrographic analysis to determine quantitatively the rock mineralogy, which can be used to assess the degradation potential.

Compactibility, dry density and water content
An important consideration in the selection of material for use as fill is its compactibility [C5.3.2(3)P], i.e. whether it can be compacted to the necessary dry density, air voids or other state so as to satisfy the design requirements. The compactibility of fill depends on the type of material and its water content. A compaction test, along with other relevant tests on the compacted material, should be carried out to check its compactibility.

The natural water content is generally not relevant in the selection of rock or granular materials for use as fill as the water content of these can be altered by drainage or, if necessary, by the addition of water to facilitate compaction. This is not possible, however, with cohesive soil with a high water content as the only options available for reducing the water content of such material are to mix it with soil having a lower water content or to dry the material. Suitable dry soil may not be readily available and the climate may not be suitable for drying out the fill within the time constraints of a project. Therefore the in situ water content is an important factor that needs to be considered when selecting a cohesive material for use as fill.

Cohesive fill material is liable to swell or collapse, depending on its dry density, its compacted state and the long-term variations in water content. The prediction of these movements is an underdeveloped area in geotechnical engineering, hence the caution in C5.3.2(2) in permitting the use cohesive materials as structural fill.

5.2.4 Compaction and Selection Criteria
The third step in the design of fill is choosing the compaction and selection criteria. C5.3.3(1)P requires that compaction criteria be established for the selected fill to ensure that it is capable of being compacted to meet its performance requirements. These criteria can be based on a relevant strength test, on achieving a minimum dry density compared with a reference dry density, or a maximum air voids. The criteria can also be based on the results of tests, such as permeability or durability tests. Selection criteria shall be chosen to ensure that selected material can be compacted to meet the compaction criteria. For cohesive material, these selection criteria can be based on the water content or Moisture Condition Value (MCV).

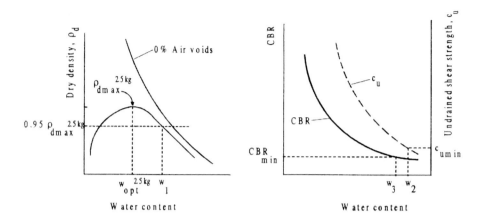

Figure 5.2: *Dry density, c_u and CBR versus water content*

Compaction criteria

The compaction criteria for cohesive and granular fill are developed from laboratory compaction tests, together with other relevant tests on the compacted material, or from the results of field trials carried out using the compaction energy level to be used during construction. An example of a laboratory compaction curve for a typical cohesive material, obtained from a Standard Proctor compaction test, with a 2.5kg rammer falling 0.3m onto the material, is shown in Figure 5.2. The maximum dry density for this test is $\rho_{dmax}^{2.5kg}$ and the optimum water content is $w_{opt}^{2.5kg}$.

Recommendations regarding the dry density to be achieved in a structural fill are given in **C5.3.4(2)** which requires that structural fill shall comprise suitable material with an average dry density equal to 100% of $\rho_{dmax}^{2.5kg}$ and a limit on the minimum dry density, ρ_{dmin} equal to 97% of this value, as shown in Table 5.2. These recommendations are based on the general experience that normal light structures have performed satisfactorily when placed on granular fill compacted to these criteria. The deformations of cohesive fill compacted to these criteria would be greater than the deformations of granular fill, hence, as noted in Table 5.2, this clause also requires that the risk of collapse and differential settlements shall be prevented.

Collapse and differential settlements may occur in cohesive soil as a result of volume changes, including swelling. These volume changes result from reductions in suction pressures arising from long term moisture movements in fill compacted dry of w_{opt} or due to dissipation of water pressures in fill compacted wet of w_{opt}. For certain design situations, it may be possible to keep these movements within acceptable limits provided the ρ_{dmin} and maximum air voids of the cohesive fill do not exceed certain specified limits. Collapse and differential settlements may occur in rock fill due breakage of the particles as a result of high contact stresses or due to softening of the rock over time. Hence it is necessary to ensure that the rock

Compaction criteria*	Average dry density	Minimum dry density
Structural fill*	*100% of $\rho_{dmax}^{2.5kg}$	*97% of $\rho_{dmax}^{2.5kg}$
Road embankment fill*	*95% of $\rho_{dmax}^{2.5kg}$ (Not an EC7 criterion)	-
* An additional requirement in the case of these compaction criteria is that the risk of collapse and differential settlements shall be prevented [C5.3.4(2)]		

Table 5.2: *Dry density compaction criteria for fills*

particles have the necessary strength and durability under the design conditions.

The recommendations in **C5.3.4(2)** regarding the dry density to be achieved in a structural fill relate to the degree of compaction in the Standard Proctor compaction test. A higher level of compaction, however, may be more appropriate for the design of structural fill to support structures with a higher than normal sensitivity to differential settlements. **C3.3.5(2)** mentions that the Standard Proctor and the Modified Proctor (4.5kg rammer) tests are the most frequently used compaction tests and correspond to different standard energy levels of compaction. The procedures for these tests are not detailed in either Part 1 or 2 of EC7 but **C10.2** of Part 2 provides the essential requirements for compaction tests. **Annex B.10** of Part 2 refers to Clause 3 of BS 1377 (1990), Part 4, which gives the detailed testing procedures for the Standard and Modified Proctor tests as well as the procedure for the compaction test using the vibrating rammer. It is also worth noting that **C3.3.5(1)P** defines the degree of compaction as the ratio between the dry unit weight of the compacted fill and the maximum dry unit weight obtained from a standard compaction test. Thus the degree of compaction can be based on any standard compaction test and is not uniquely defined in terms of the Standard Proctor test.

In the case of rock fill and some coarse granular fill material, the size of the particles mean that the Proctor compaction test is not applicable and hence is difficult to establish practical compaction criteria for such material. Instead the compaction of such material can be checked as outlined in §5.2.5 below.

Cohesive fill is used extensively in road construction but, as fill for road embankments is not explicitly covered by EC7, no compaction criteria are provided for this situation. However it has been found, from experience, that the deformations of cohesive fills for roads compacted to an average dry density equal to 95% of $\rho_{dmax}^{2.5kg}$ are generally acceptable and hence the serviceability limit state requirements are satisfied. This compaction criterion is also shown in Table 5.2.

Selection criteria

The water content, w_1 in Figure 5.2 is the upper limit of the water content at which the material, whose compaction curve is plotted in Figure 5.2, can be compacted to achieve the criterion of 95% of $\rho_{dmax}^{2.5kg}$. This upper limit on the water content, therefore, is a selection criterion for suitable cohesive material whose value depends on the compaction criterion required to satisfy the design criteria for a road to be placed on the fill.

Where w_1 is the upper limit on the acceptable water content and the natural material is found to have a water content in excess of this value, it will either have to be mixed with drier soil or else dried out, as permitted in **C5.3.2(4)**. If it is not feasible for the water content to be reduced within the constraints of the project, then the soil is unsuitable as it cannot be compacted to a high enough dry density so as to satisfy the fundamental requirement regarding the deformation of the fill.

Some compaction criteria may require an upper limit to be placed on the water content of the selected material; for example, the undrained shear strength of cohesive fill when compacted at w_1 may give rise to unstable side slopes. Should that be the case, then the water content w_2 should be determined which would provide a higher shear strength, as shown in Figure 5.2, and hence ensure that the level of safety against shear failure of the side slopes is satisfactory.

A further requirement, known as trafficability, is that the compacted fill is able to support construction traffic [**C5.3.1(2)**]. This requirement is normally assessed by measuring the penetration resistance in a California Bearing Ratio (CBR) test and checking that the CBR of the material, when compacted, is not less than a certain minimum value, CBR_{min}. This would give a further upper limit on the water content of the selected cohesive material, say w_3. Thus the design upper limit of the water content of a cohesive fill to meet the fundamental requirements for a road embankment should be the lower of w_1, w_2, and w_3. It may also be necessary to place a lower limit on the acceptable water content of cohesive material because, if such soils are too dry, it may not be possible to achieve an adequate dry density with a maximum allowable air voids using normal compaction equipment.

In practice the spatial variability of cohesive material normally makes the use of upper water content limits linked directly to the required ρ_{dmin}, c_u or CBR values, impractical. Instead it is more usual to relate the upper water content limit to the w_{opt} value for the level of compaction used. For example, the compaction criterion may be expressed as an upper limit on the water content equal to $1.2 * w_{opt}^{2.5kg}$.

As an alternative to measuring the water content of a material, its suitability as fill may be assessed on the basis of other parameters measured in tests such as the CBR and MCV tests (BS 1377, 1990). For example a CBR value of 2% may be chosen as the selection criterion if this has been shown by trial to give a cohesive fill which satisfies the fundamental requirements. The use of the MCV, which is a parameter that defines the limit of suitable material, is common in some countries. The MCV is an empirical parameter whose value can be related to a required minimum dry density or a minimum strength of a cohesive fill.

5.2.5 Fill Placement and Compaction Procedures

The fourth step in the design of fill is selection of the appropriate placement and compaction procedures. Inclusion of the need to consider the construction procedures for fill placement and compaction as part of the design of fills reflects the close dependence of the quality of the final fill on the construction method. **C5.3.3(2)P** requires that the fill placement and compaction procedures be selected so as to satisfy the compaction criteria discussed in §5.2.4 above and hence ensure the fill is stable and natural subsoil is not adversely affected. The actual procedures to be adopted will depend on the compaction criteria, on the nature of the material and on the compaction equipment to be used. **C5.3.3(4)** recommends that a trial field test should be made with the intended fill material to develop an appropriate procedure for compaction. This is important as it ensures that the required performance requirements can be achieved on site with the selected compaction equipment and compaction procedures. The trial test may also be used to select the criteria for checking the compacted fill. A trial test may not be required for small to medium GC1 projects, as **C5.3.3(4)** is an Application Rule, not a Principle.

Overcompaction, which can cause problems due to the formation of shear surfaces (slickensides), high lateral earth pressures on buried structured or crushing of particles, is to be avoided should these be not acceptable [**C5.3.4(3)P**].

When placing cohesive fill, the fill surface shall be shaped to give good drainage [**5.3.3(5)P**], and frozen, expansive or soluble soils shall not be used [**C5.3.3(6)P**]. The fill may require heating and its surface may require some form of frost protection if it placed in temperatures below freezing point [**C5.3.5(7)**].

5.2.6 Checking Fill

The fifth step in the design of fill is selecting a system for checking that the compacted fill meets the design requirements. This is an example of the supervision of construction covered in **C4.4** and §4.2.4. **C5.3.4(1)P** requires that the compaction of fill be checked either by inspection or by testing. These would normally be linked to a method type or an end product type specification respectively, as described below.

Method specification

When using method specification, the compaction of the fill may be checked by inspecting (controlling) the compaction procedure to ensure that it is carried out according to a specified method. This requires that material to be used be first checked to ensure that it can be compacted so as to satisfy the fundamental requirements. The compaction method specification should include requirements for the fill placement and compaction procedures, including the compaction equipment, the layer thickness and the number of passes of the equipment. The requirements for the fill placement and compaction procedure should be selected so that the compacted fill meets the fundamental requirements and should be based either on a field trial or on comparable experience.

End product specification

In the case of end product specification, the required characteristics of the compacted fill are specified and the compacted fill is tested to check that these have been achieved. Testing the compacted fill can involve measuring the dry density, water content, penetration resistance, stiffness, etc. Interpretation of the measurements should ensure that the relevant property, for example the required dry density or CBR, is obtained for the compacted fill. Measurement of the penetration resistance (as in a CBR test) or the stiffness of a fill (as in a plate loading test) may be used as compaction criteria. However EC7 warns that such measurements may not indicate whether adequate compaction has been achieved in cohesive soils [**C5.3.4(2)**]. This relates to the high undrained strengths and stiffnesses which can be achieved in dry cohesive soils from suction pressures even when the material is poorly compacted. This strength is lost with an increase in water content and can lead to collapse of the soil structure and excessive settlement.

Checking rock fill

The size of particles in rock fill can make it difficult to specify practical compaction criteria and site control measures for such material. The normal sand replacement or nuclear density meter methods for measuring density do not always operate successfully in rockfill material, which can have relatively large voids. This dilemma is recognised in EC7 which includes the pragmatic clause **C5.3.4(2)** for rock fill and for fill containing large amounts of coarse particles, for which the Proctor compaction test is not applicable. This clause notes that the compaction of such fill can be checked by one of the following methods:

* checking compaction has been performed in accordance with procedures developed from a field trial or from comparable experience;
* checking that additional settlement induced by an additional pass of the compaction equipment is lower than a specified value; or
* using a seismic method to check the density.

5.3   Dewatering

Dewatering covers the abstraction of water both to improve the properties of ground and to facilitate construction. In some situations groundwater recharge may also be required as part of a dewatering scheme to prevent drawdown causing unacceptable settlements of neighbouring structures. The conditions to be satisfied and the aspects to be checked when designing a dewatering scheme to EC7 are set out in **C5.4**. These items are summarised in the checklist in Table 5.3. The observational method (**C2.7**) can be used for the design of dewatering schemes.

5.4   Ground Improvement and Reinforcement

The design of ground improvement and reinforcement schemes is only covered in general terms in EC7. In many cases such schemes would be classified as GC3 [**C5.5(3)**] and, as such, would not be fully covered by the code requirements in EC7 and should be carried out by a geotechnical specialist. **C5.5(2)** lists some of the aspects which should be considered when designing a ground improvement or ground reinforcement scheme. This list includes, for example, the thickness of the in situ strata or fill material, the nature, size and position of the structure to be supported by the ground, the prevention of damage to adjacent structures, etc. Before any ground improvement scheme is chosen or used, **C5.5(4)P** requires that the effectiveness of ground improvement be checked against the acceptance criteria by determining the changes in the appropriate ground properties or ground conditions resulting from the improvement method. So that this can be done, **C5.5(1)P** requires that a geotechnical investigation be carried out before any ground improvement to determine the initial conditions. Guidance on the execution of ground improvement and ground reinforcement methods can be obtained from the following European Standards prepared by the CEN committee TC 288:

* EN 12715 Execution of special geotechnical works – Grouting;
* EN 12716 Execution of special geotechnical works – Jet grouting.

Items to be checked	Checked
Sides and base of excavations remain stable and heave or rupture of base due to excessive water pressures beneath less permeable layers does not occur.	
Settlements or damage to nearby structures due to the dewatering scheme is not excessive.	
Excessive loss of ground due to erosion by seepage is avoided.	
Adequate filters are provided around sumps, where necessary.	
Water is discharged well clear of excavations.	
Water levels and pore water pressures are adequately monitored during dewatering.	
Interpretation procedures for reviewing water level and pore water pressure data after installation are adequate.	
Dewatering scheme is able to keep fluctuations in water levels and pore water pressures within acceptable limits.	
An adequate margin in pumping capacity is provided in case of a breakdown.	
Ground and structural movements arising from return of the groundwater to final equilibrium do not cause problems, such as the collapse of loose sands.	
The scheme does lead to the excessive transport of contaminated water to the excavation.	
The scheme does not lead to excessive abstraction of drinking water from the surrounding area.	
The long term corrosion and clogging of the dewatering system has been considered, where relevant.	

Table 5.3: *Items to be checked when designing a dewatering scheme to EC7*

TC 288 is preparing a standard for the execution of reinforced soil at present and there are proposals to prepare execution standards for soil improvement and dewatering. These execution standards normally do not, and if they do, should not, contain design rules, which should be provided in EC7. It is intended that any design rules in these standards will be transferred to EC7 when it is being revised.

References

BS 1377 (1990) *Methods of test for soils for civil engineering purposes - Part 4: Compaction related tests*, British Standards Institution (BSI), London.

BS 6349 (1984) *Code of practice for maritime structures*, British Standards Institution, London.

BS 812: Part 113 (1990) *Testing aggregates - Part 113: Aggregate abrasion value test*, British Standards Institution, London.

ISRM (1981) *Rock characterisation, testing and monitoring, ISRM suggested methods*, ed. E T Brown, Pergamon Press.

Leps, T.M. (1973) Flow through rockfill, *Embankment dam engineering*, Wiley-Interscience.

Chapter 6

Spread Foundations

6.1 Introduction

Section 6 of EC7 is mainly concerned with the design of spread foundations, such as pad, strip and raft foundations, often referred to as shallow foundations. Spread foundations are foundations at shallow depths of burial where the ground resistance on the sides of the foundation does not contribute significantly to the bearing resistance. Some of the provisions in this Section may also apply to deep foundations, such as caissons and piers [**C6.1(1)P**].

Although commonly used in practice, the term "shallow foundation" has not been used in EC7 because of the difficulty in defining this term. One definition of a shallow foundation is when the founding depth (D) is less than the width (B), i.e. D/B < 1. This definition may be acceptable for pad and strip foundations but does not apply in the case of a raft foundation. Another definition sometimes used for a shallow foundation is when the founding depth is less than 3m.

The design of spread foundations using the limit state method and partial factors is described in this Chapter. The influence of the partial factors on the bearing resistance and the margins of safety for spread foundations are examined. A number of worked examples show how spread foundations are designed to EC7.

6.2 Limit States for Spread Foundations

The first requirement in EC7 regarding the design of spread foundations is that a list of the ultimate and serviceability limit states to be considered shall be compiled [**C6.2(1)P**]. Spread foundations need be checked against failure due to the exceedence of any of these limit states.

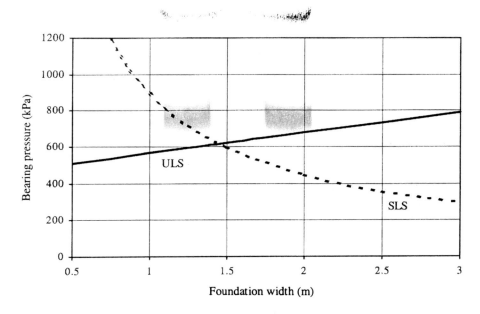

Figure 6.1: *Bearing pressures versus ULS and SLS design foundation widths*

Ultimate Limit States
The most important ULS is usually collapse due to bearing resistance failure, however the possibility of the following other ULSs also needs to be considered:
- loss of overall stability (e.g. a slope stability type of failure);
- failure by sliding;
- combined failure in the ground and the structure;
- structural failure due to excessive foundation movements.

Serviceability Limit States
Serviceability limit states include movements and deflections which damage the appearance of the supported structure, reduce its useful life or cause damage to finishes, non-structural elements, machinery or other installations in the structure. The movements and deflections which need to be considered in the case of spread foundations include:
- excessive settlements, both total and differential;
- excessive rotations;
- excessive heave;
- excessive vibrations.

In some situations the ULS controls the design of spread foundations while in other situations the SLS is the controlling condition. For example, in the case of a foundation supporting a light load, the design foundation width is narrow and the controlling condition is the ultimate limit state due to localised failure beneath the foundation caused by high bearing pressure. In the case of a foundation supporting a large load, however, the design foundation width is greater and the controlling condition may be the SLS due to excessive settlement of the foundation resulting from the wider foundation loading a larger volume of soil.

The variation in the limit state which controls the design with regard to bearing pressure is illustrated by the graphs in Figures 6.1. These show the bearing pressures (characteristic load/ foundation area) plotted against the ULS and SLS design foundation width for a square pad foundation resting on sand with $\phi'_k = 35°$ and a constant $E'_m = 25MPa$. These graphs were obtained using Equations 6.5 and 6.12 for the bearing resistance and settlement, assuming a maximum acceptable settlement, $s_{max} = 25mm$ and $f = 0.6$ (see §6.6.1). The graphs show that for small foundation widths, the design width is controlled by the ULS, while for larger widths, the design is controlled by the SLS. This is because wider foundations increase the stresses in a larger volume of ground causing more settlement, resulting in the SLS becoming the controlling condition.

6.3   Actions and Design Situations

Apart from permanent and variable loads from structures, the more important of the actions in C2.4.2(4)P to be considered in the design of spread foundations are those due to:

- water pressures;
- the removal of load or excavation of the ground;
- swelling and shrinkage caused by vegetation, climate or moisture changes;
- movements due to creeping or sliding soil masses; and
- movements due to self-compaction.

Some advice on how to design foundations against swelling or shrinking due to vegetation, climate or moisture changes is given in §6.4.

For perfectly flexible, i.e. statically determinate, structures the vertical loading is distributed directly to the foundations without any redistribution of the loading from one part of the structure to another. This implies that differential settlements can occur. In the case of stiff, indeterminate structures, ground-structure interaction may redistribute the forces in the structure and this may need to be analysed to determine the forces on the foundations [C6.3(2)]. Generally this redistribution in stiff structures is favourable with regard to the geotechnical design because the forces on the foundations are increased where the ground is strongest, and hence most able to support them, and reduced where ground is weakest. Reducing the stiffness of a structure, while increasing the differential settlements, reduces the maximum foundation forces and bending moments and so may result in a more economical design. Those structures where consideration of ground-structure interaction is most likely to be important and where reducing the stiffness may result in more economical designs include raft foundations and structures that are relatively stiff compared to the ground.

The design situations that need to be considered are listed in **C2.2(1)P**. In the design of spread foundations it is most important to assess the depth of the groundwater table. For example, if the groundwater table rises to the ground surface, the drained bearing resistance is reduced to about half the value when the groundwater level is at great depth. This is because the q' and γ' terms in Equation 6.5 both involve the effective unit weight of the soil $\gamma' = \gamma - \gamma_w$, where γ is the total soil unit weight and γ_w is the unit weight of water; q' involves the effective soil weight above the foundation level and γ' the effective soil weight below the foundation level. As γ_w is about half γ, q' and γ' both reduce by about one half when the groundwater table rises from great depth to the surface.

6.4   Design and Construction Considerations

6.4.1 Items to be Considered

EC7 lists in **C6.4(1)P** a number of items that need to be considered when choosing the depth and location of a spread foundation. These items include the following:

- *Reaching an adequate bearing stratum.* The ground at the bottom of excavations for spread foundations should always be inspected to check that an adequate stratum has indeed been reached.
- *The depth above which shrinkage and swelling of clay soils, due to seasonal weather changes, or to trees and shrubs, may cause appreciable ground movement.* Whenever possible foundations should not be placed close to trees or, on clay soils, close to where trees have recently been growing. Where the penetration of tree roots is likely to cause damage due to settlement, heave or uplift, foundations should be placed at a sufficient depth so as to avoid such damage.
- *The distance from trees.* In addition to placing foundations at sufficient depth, it is also important to place foundations at a sufficient distance from trees. In the case of highly shrinkable clay soils, the Building Research Establishment (BRE, 1987) recommends that, to avoid damage due to subsidence or heave caused by tree roots, foundations should normally be separated from the trees by a distance of at least 0.5H, where H is the mature height of the trees. Some trees have very high moisture requirements and consequently are more likely to cause damage to buildings than other species. The root systems of trees such as the poplar, oak, elm, willow and eucalyptus are particularly liable to cause settlement of foundations and, in the case of these trees on highly shrinkable clays, it is recommended that the separation distance be doubled to H. Other trees that can cause damage at distances greater than 0.5H include cherry, plum, whitebeam and rowan trees. When foundations are placed at a depth of 1m and separated from trees by the distances recommend above, the risk of damage due to the influence of trees is likely to be low.

- *The depth above which frost damage may occur.* Foundations need to be founded deeply enough to protect them from damage due to frost. The depth above which frost damage may occur depends on the local climate.
- *The level of the water table in the ground and the problems that may occur if excavation for a foundation is required below this level.* The possibility that excavations below the water table may result in softening of cohesive soils or the collapse of excavation sides in non-cohesive soils should be considered. If pumping (dewatering) is used to keep an excavation dry, the possibility that this may cause drawdown of the water table and settlement of neighbouring buildings resting on compressible soils should be investigated.
- *Possible ground movements and reductions in the strength of the bearing stratum by seepage or climatic effects or by construction procedures.* It is important to ensure that, after excavation and before placing a foundation, the bearing stratum is not weakened by water or by construction activity.

  Blinding concrete: According to **C4.1.3.3(111)** of EC2-3, a blinding layer of concrete may be used to protect the bottom of an excavation for a foundation. This blinding layer should be placed as soon as possible after trimming the excavation to the required depth. It should consist of lean-mix concrete about 50 to 75mm thick. Its purpose is to cover the soil so as to avoid contamination, and protect the foundation soil from the effects of wetting, drying or frost, and to maintain a flat surface for the structural concrete forming the foundation. **C6.3.5.1.2(101)** of EC2-3 recommends that this blinding concrete be compacted and finished smooth with a vibrating plate or trowel.
- *The effects of excavations required for construction of nearby foundations and structures and future excavations for services close to the foundation.* The effects of excavations for neighbouring structures and services is an important factor to be considered when choosing the depth of a foundation because of the effect that such excavations can have on the stability of foundations, particularly foundations resting on non-cohesive soil.
- *Practical considerations related to economic excavation, setting out tolerances, working space requirements and the dimensions of the wall or column supported by the foundation.* When selecting foundation widths, allowance should be made for the fact that excavations for foundations in some soils cannot be constructed with great accuracy. According to C6.2.2.1 of EC2-3, the permissible geometric deviations should be reasonably large, bearing in mind the nature of the construction. Particular care is required in the case of narrow strip foundations as large errors in the setting out of trenches for strip foundations may result in excessive eccentricities of the loading on foundations and possible failure or tilting of foundations. For foundations close to the boundary of a site, the geometric tolerances should be specified such that the foundations do not project beyond the site boundary.

6.4.2 Design Methods and Calculation Process

Direct and Indirect Design Methods
Two methods may be used to design spread foundations. The first method is the direct method in which separate calculations are carried out to check against the exceedence of a ULS and a SLS. When checking a ULS, the calculation shall model the failure mechanism as closely as possible, while a deformation analysis is used to check a SLS [**C6.4(3)P**]. An indirect method is where, using SLS loads, one calculation is carried out, or a bearing pressure is chosen, to avoid the occurrence of a SLS and, at the same time, to avoid the occurrence of all other possible limit states. Presumed bearing pressures are an example of this method for GC1, as discussed in §2.6. Further information about indirect methods is provided in §6.6.2.

Calculation Process
The general calculation process for designing the width, B of a spread foundation to EC7 is shown by the flow diagram in Figure 6.2 where the parameters V_d, R_d, E_d and C_d are defined in §6.5.2 and §6.6.1. This process involves checking the ULS and SLS. The ULS should be checked for both Cases B and C and for undrained conditions, in the case of fine-grained soils (usually referred to as cohesive soils in EC7), and drained conditions. As explained in §6.5.3, Case C generally controls the size of foundations for the ULS, except for certain loading and soil combinations, such as foundations with highly eccentric loading on non-cohesive soils as shown in §6.5.4 and by Example 6.4.

6.5 Ultimate Limit State Design

6.5.1 Loss of Overall Stability

Loss of overall stability is a failure mechanism involving the whole mass of ground containing a foundation. Such a failure mechanism is most likely to occur in the case of a foundation in sloping ground or close to an excavation or river. The design procedures for checking against loss of overall stability are covered in Chapter 9 on Embankments and Slopes.

6.5.2 Bearing Resistance Failure

When designing a foundation to EC7 it is necessary to check that the foundation has adequate safety against bearing resistance failure by demonstrating that the ULS design action, V_d does not exceed the ULS design bearing resistance, R_d:

$$V_d \leq R_d \qquad (6.1)$$

As the design values V_d and R_d include partial factors on the loads and material properties, no other safety factors are applied when using Inequality 6.1; it is only necessary to ensure that V_d does not exceed R_d.

It should be noted that the term *bearing resistance* is used in EC7 in place of the traditional term *bearing capacity*. This is to make EC7 consistent with the other

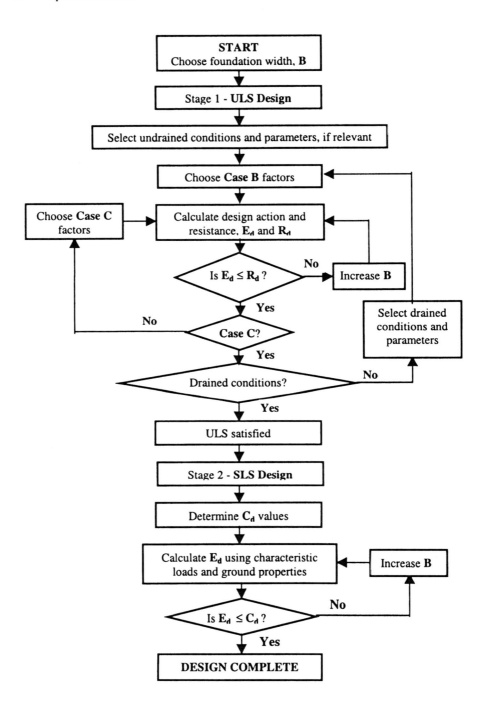

Figure 6.2: *General calculation process for spread foundation designs to EC7*

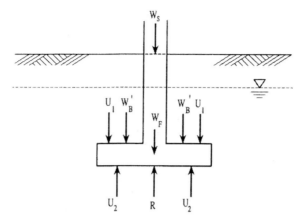

Figure 6.3: *Design actions and resistance on a spread foundation*

Eurocodes which involve the resistance of other materials and because the term capacity is used in the Eurocodes for something having the units of volume. The advantage of using the term bearing resistance rather than bearing capacity is that it is more logical and has a clearer meaning. It should also be noted that, while in traditional calculations the bearing capacity of a spread foundation is a pressure with units kPa, in EC7 the bearing resistance, R_d is a force, with units kN. The traditional concept of an allowable bearing pressure is not used in EC7 because it is not well defined and generally does not take account of the geometry of the foundation or the nature of the loading. Presumed bearing pressures, which have some similarities with allowable bearing pressures, are discussed in §2.6.

ULS Design Action
The ULS design action, V_d is the force normal to the foundation (usually the vertical component of the load) and includes the load from the structure W_S, the weight of the foundation W_F, and the weight of any backfill material W'_B.

For drained conditions, the water pressures are known quantities, and therefore, in accordance with the definition of actions in **C2.4.2(1)P**, water pressures forces are generally included as actions in V_d **[C6.5.2.1(1)P]** Hence, in the situation illustrated in Figure 6.3 of a foundation beneath the groundwater level, the effective weight of the backfill material, W'_B and the force due to the water pressure on top of the foundation, U_1 is treated separately and added to V_d, while the uplift force, U_2 due to the water pressure on the underside of the foundation is subtracted from V_d so that:

$$V_d = W_S + W_F + W'_B + U_1 - U_2 \qquad (6.2)$$

Where the water pressure is hydrostatic, this is equivalent to using the buoyant (submerged) weight of the foundation and the effective weight of the backfill material below the groundwater level **[C6.5.2.1(2)]**, which is the method used in traditional drained analyses.

For undrained conditions, the water pressures are not known quantities, or are not determined, and so are not included in V_d as actions. V_d is the total ULS design load normal to the foundation due to the total weight of everything above the foundation. If the foundation lies below the groundwater level, the foundation is still subjected to the effect of buoyancy due to the uplift force of the water pressure on the base of the foundation. However this reduction in the foundation load is equal to the reduction in the bearing resistance due to the submerged weight of the ground beside the foundation. These cancel each other and hence the uplift force due to the water pressure is omitted from V_d when carrying out a total stress analysis for undrained conditions. Example 6.1 shows how V_d is calculated for drained and undrained conditions.

ULS Design Bearing Resistance

C6.5.2.2(1)P requires that the ULS bearing resistance be checked for both the short-term and long-term situations, i.e. for undrained conditions using c_u and for drained conditions using c' and φ'. This is particularly important for fine-grained soils, such as silts and clays, where changes in pore water pressure may lead to changes in shear strength. Undrained conditions are only appropriate for short term conditions during which the excess pore water pressures due to the loading do not have time to dissipate. For long term conditions, foundations must be designed using the effective stress strength parameters, c' and φ'. The selected shear strength parameters need to take account of any structural characteristics of the ground [**C6.5.2.2(3)P**], such as layering. For example, if a strong formation underlies a weak formation beneath the foundation, the relevant bearing resistance may be calculated using the shear parameters of the weak formation [**C6.5.2.2(5)**].

Annex B of EC7 provides equations for the vertical bearing resistance, together with equations for the bearing resistance factors and the dimensionless factors for the effects of foundation shape and inclination of the load. In these equations it is assumed that a linear (meaning *uniform*) bearing pressure acts on the design effective foundation area, A'. The effective foundation area is defined as the foundation base area or, in the case of an eccentric load, as the reduced foundation base area whose centroid is the point through which the load resultant acts, and is equal to B'L' (the effective width times the effective length). It should be noted that this approach assumes a constant bearing pressure, equal to the failure pressure, on the effective base of the foundation. This differs from the traditional approach where, in the case of an eccentric load, the bearing pressure is assumed to vary linearly across the full width of the foundation and the bearing pressure is only at its failure value at the edge of the foundation.

The determination of B' is illustrated in Figure 6.4 by the example of a rectangular pad foundation of breadth, B and length, L supporting a design vertical central load, V and a design horizontal load, H acting parallel to the width and at a height, h above the foundation base. The eccentricity, e of the resulting force on the foundation is obtained by taking moments of the forces about the foundation centre-line and is equal to the distance Hh/V from the centre-line. Hence the effective foundation width is:

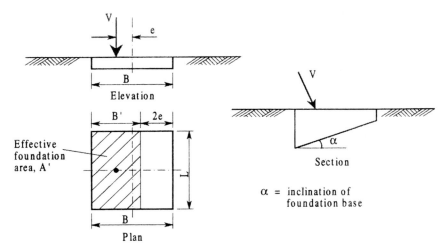

Figure 6.4: *Effective foundation area and base inclination angle*

$$B' = 2(B/2 - e) = B - 2Hh/V \qquad (6.3)$$

The effective foundation length, L' in this example is equal to L.

Undrained Resistance

For undrained conditions, the equation in EC7 for the design bearing resistance is:

$$R_d/A' = (2 + \pi) c_u s_c i_c + q = 5.14 c_u s_c i_c + q \qquad (6.4)$$

with q as the design total overburden pressure at the level of the foundation base and the design values of the dimensionless factors for:

- the inclination of the load due to a horizontal load H $i_c = 0.5\left(1 + \sqrt{1 - \dfrac{H}{A'c_u}}\right)$

- the shape of the foundation: $s_c = 1 + 0.2(B'/L')$ for a rectangular shape
    $s_c = 1.2$ for a square or circular shape

Factors b_c and d_c for the inclination of the base and depth of embedment are not included in EC7 because of concern about their reliability. However, it is likely the following equations for these factors, taken from DIN 4017 (1979a), will be included in the EN version and applied to the cohesion term of Equation 6.4:

$$b_c = 1 - 2\alpha/(2 + \pi) \text{ where } \alpha \text{ is shown in Figure 6.4}$$

$$d_c = d_q = 1 + 0.35(D/B') \text{ and } d_c \le 1.7$$

The slope inclination factor, b_c is not normally used in GC2 designs. The d_c value is greater than unity and takes account of the increase in bearing resistance provided by the strength of the soil above the foundation base. As this soil may be weak or disturbed, d_c should only be used if the strength of the soil above the foundation base is not significantly less than the strength below the base. By not using d_c, i.e. by assuming this value is equal to unity when calculating the bearing

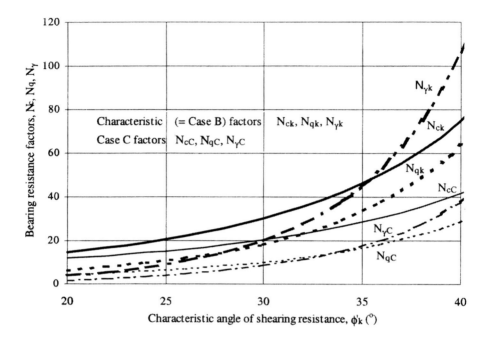

Figure 6.5: *Bearing resistance factors*

resistance, EC7 has adopted a conservative approach that provides a lower value for R_d. This approach has been adopted in Examples 6.1 to 6.5.

Drained Resistance

For drained conditions, the equation in EC7 for the design bearing resistance is:

$$R_d/A' = c'N_c\, s_c\, i_c + q'N_q\, s_q\, i_q + 0.5\gamma'B'N_\gamma\, s_\gamma\, i_\gamma \qquad (6.5)$$

with q' as the design effective overburden pressure at the foundation level base and the design values of the dimensionless factors for:

- bearing resistance: $N_q = e^{\pi \tan \phi'} \tan^2(45 + \phi'/2)$

  $N_c = (N_q - 1) \cot\phi'$

  $N_\gamma = 2(N_q - 1) \tan\phi'$ when $\delta > \phi'/2$ (rough base)

- the shape of the foundation:

  $s_q = 1 + (B'/L')\sin\phi'$ for a rectangular shape

  $s_q = 1 + \sin\phi'$ for a square or circular shape

  $s_\gamma = 1 - 0.3(B'/L')$ for a rectangular shape

  $s_\gamma = 0.7$ for a square or circular shape

  $s_c = (s_q N_q - 1)/(N_q - 1)$ for rectangular, square or circular shape

- the inclination of the load, caused by a horizontal load H:

$$i_c = i_q - (1 - i_q)/N_c\tan\phi'$$

$$i_q = [1 - H/(V + A'c'\cot\phi')]^m$$

$$i_\gamma = [1 - H/(V + A'c'\cot\phi')]^{m+1}$$

with $m = m_B = [2 + (B'/L')]/[1+(B'/L')]$ when H acts in the direction of B'

and $m = m_L = [2 + (L'/B')]/[1+(L'/B')]$ when H acts in the direction of L'.

The values of the bearing resistance factors N_q, N_c and N_γ calculated using the characteristic and design ϕ' values, i.e. for Cases B and C, are presented in Figure 6.5 as the graphs N_{qk}, N_{ck} and $N_{\gamma k}$ and N_{qC}, N_{cC} and $N_{\gamma C}$ plotted against ϕ'_k.

It should be noted that the equations given in EC7 for the inclination factors for a horizontal load parallel to B' and L' are incorrect as they lead to two different sets of values when B' is equal to L'. For this reason the values in EC7 have not been given above; instead inclination factors from DIN 4017 (1979b) are presented which do not have this problem. These corrected values are used in the subsequent calculations and will probably be included in the EN version of EC7.

As in the case of the undrained resistance, factors b and d for the inclination of the base and depth of embedment are not included in Equation 6.5. However, the following factors, taken from DIN 4017 (1979a), are likely to be included in the EN version of EC7 and applied to the c, q and γ components of Equation 6.5:

$$b_c = b_q - (1 - b_q)/(N_c\tan\phi')$$

$$b_q = b_\gamma = (1 - \alpha\tan\phi')^2$$

$$d_c = d_q = 1 + 0.35(D/B') \text{ and } d_c \leq 1.7$$

$$d_\gamma = 1.0$$

For the reasons given above, these factors are not provided in EC7 and are not used in the examples in §6.9.

6.5.3 Safety Margins

Since partial factors are applied to both the loads and resistances in ULS designs to EC7, the overall safety margin or factor of safety (FOS) against bearing resistance failure may be assessed from the following equation:

$$FOS = \left(\frac{E_d}{E_k}\right)\left(\frac{R_k}{R_d}\right) \qquad (6.6)$$

where E_d, E_k, R_d, R_k are the design and characteristic (unfactored) action effects and resistances, respectively. Using the partial factors in Table 2.6, this equation shows that, for vertical loading and undrained conditions, the FOS values for Case B range from 1.35 to 1.5, as $R_d = R_k$, and for Case C range from 1.4 to 1.82, depending on the proportion of permanent and variable load. This shows that Case C always has the higher FOS value and so always controls the size of the foundation for vertical loading and undrained conditions. The FOS values for Case C provide a lower safety margin for foundations for undrained conditions than that provided by the typical overall safety factors of 2.5 to 3 normally used in traditional designs, particularly when the design load is mainly permanent. This is

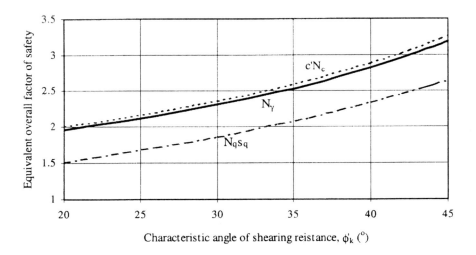

Figure 6.6: *Variation in ratio of characteristic and design bearing resistance components*

because the safety factors in traditional designs are also intended to ensure that settlements are not excessive and hence serviceability limit states do not occur.

For drained conditions, the FOS values for Case B for vertical loading are the same as for undrained conditions and range from 1.35 to 1.5, depending on the proportion of permanent and variable load. The ratios of the characteristic and Case C design values of the different bearing resistance components in Equation 6.5 for a square foundation are plotted against ϕ'_k in Figure 6.6. These graphs show that the $c'N_c$ and N_γ ratios range from about 2 for $\phi'_k = 20°$ to over 3 when $\phi'_k = 45°$ while the $N_q s_q$ ratios are lower, ranging from about 1.5 to 2.5. These values are all greater than the Case B values and show that Case C always controls the ULS design for vertical loading for drained conditions also. If some of the loading is variable, these FOS values will be increased by up to a maximum of 30% and hence will result in FOS values similar to the values of 2.5 to 3 used in traditional designs. The lower $N_q s_q$ ratios indicate that caution may be required when the overburden pressure is a large component of the bearing resistance as excessive settlements may occur due to the lower safety margin. While Case C always controls the size of spread foundations subjected to vertical loading with regard to the ULS, Case B may control the size under certain loading and ground conditions, for example a highly eccentrically loaded foundation on non-cohesive soil as shown in §6.5.4 and by Example 6.3.

6.5.4 Eccentric Loads

Eccentric loads on foundations may be caused by either non-central vertical loads or by horizontal loads, for example by wind or earth pressure. The way these are treated in designs to EC7 is to assume the bearing resistance is provided by the

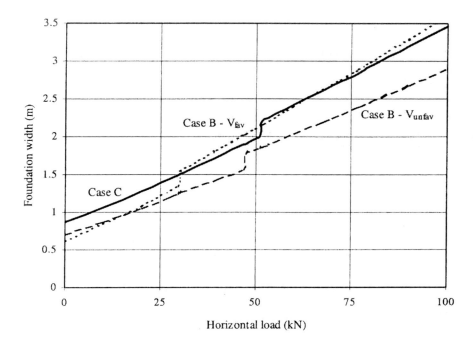

Figure 6.7: *Case B and C widths for an eccentrically loaded foundation*

force from a uniform pressure acting on the reduced effective area of the foundation base, A' as defined in §6.5.2. Thus the design is not based on the bearing pressure at the edge of the foundation, as is frequently the case in traditional designs.

Another difference compared with many traditional designs, is that there is no requirement to restrict the eccentricity of the load on the foundation base to one sixth of the base width so that the point of application of the load falls within the middle third. Instead **C6.5.4(1)P** requires that special precautions be taken when the eccentricity of the load exceeds 1/3 of the width of a rectangular footing or 1/6 of the radius of a circular footing. The application rule for this, **C6.5.4(2)**, recommends that differences of up to 0.1m should be considered, which means that, after calculating the foundation size for the highly eccentric loading, the dimensions of the footing should then be increased by 0.1m in all directions; i.e. the side lengths should be increased by 0.2m as shown in Example 6.3.

When designing an eccentrically loaded foundation, the vertical load is favourable when calculating the eccentricity. For Case B, therefore, the vertical permanent load is normally unfactored. However, when the horizontal load is very small, the vertical load will dominate and so factoring the vertical load, i.e. treating it as being unfavourable, results in a larger foundation (Case B - V_{unfav}), but Case C is still the controlling condition. This effect is shown by the graphs in Figure 6.7 for a square foundation, width B, at a depth of 1m in sand with $\phi'_k = 35°$.

The foundation supports a characteristic permanent vertical load, V_k of 100kN and a variable horizontal load, H_k at a height of 4m that increases from 0kN to 100kN. These graphs show that for small values of H_k, and hence small eccentricities, Case C is controls the foundation width. As H_k increases, the Case B loading with the favourable vertical load (Case B - V_{fav}) is the first to reach an eccentricity of B/3 and so B jumps by 0.2m and Case B - V_{fav} controls the design. As H_k increases further, the eccentricity of the Case C load then reaches B/3 and so the B value jumps by 0.2m and Case C controls again. As H_k increases still further, the foundation size for Case B - V_{fav} eventually exceeds that for Case C so that, at high eccentricities, Case B is the controlling condition. These changes from Case C to B as H_k increases show the importance of checking for both cases when the loading is eccentric.

6.5.5 Sliding Failure
Foundations subjected to horizontal loads should be checked for adequate safety against sliding failure[C6.5.3(1)P] by demonstrating that:

$$H_d \leq R_d \qquad (6.7)$$

where H_d is the horizontal component of the ULS design action, and includes the forces due to design active earth pressures, and R_d is the design horizontal resistance. $R_d = S_d + E_{pd}$ [C6.5.3(2)P] where S_d is the horizontal design shear resistance between the base of the foundation and the ground, and E_{pd} is the design resistance due to the passive earth pressure on the side of the foundation that can be mobilised with the displacements appropriate to the limit state and which is available during the life of the structure. For small movements it may be necessary to use only a certain proportion, e.g. 70%, of E_{pd}. For large movements, it may be necessary to use post-peak values of the strength parameters when calculating S_d and E_{pd}. [C6.5.3(3)]. EC7 warns that caution is required when using E_{pd} as the possibility of the soil in front of the foundation being removed by erosion or human activity should be considered.

Drained Resistance
For drained conditions, the drained shear resistance S_d is given by:

$$S_d = V'_d \tan\delta'_d \qquad (6.8)$$

where V'_d is the design effective vertical load on the foundation and δ' is the design friction angle on the foundation base [C6.5.3(7)P]. For cast-in-situ concrete foundations, $\delta'_d = \phi'_d$. According to C6.5.3(8), any c' should be neglected.

Undrained Resistance
For undrained conditions, the undrained shear resistance S_d is limited by:

$$S_d = A'c_u \qquad (6.9)$$

where A' is the effective foundation area as defined in §6.5.2. An additional requirement in C6.5.3(9)P in the case of clay and undrained conditions is that, if it is possible for water or air to reach the interface between the foundation and underlying clay, then:

$$S_d \leq 0.4V_d \qquad (6.10)$$

i.e. the horizontal design resistance should not exceed 40% of the vertical design load, V_d.

6.5.6 Structural Failure due to Foundation Movement

C6.5.5(1)P requires that differential settlements and horizontal movements of the foundations of a structure under the ULS loads and ground deformation parameters be considered to ensure that they do not cause a ULS in the supported structure. According to C6.5.5(2), presumed bearing pressures may be adopted such that the displacements will not cause a ULS in the structure. Swelling ground is given as an example of movement that could cause a ULS [C6.5.5(3)].

When analysing this limit state, the displacements must be calculated, for Case C, using factored (i.e. increased) loads and factored (i.e. reduced) deformation parameters. However, no ULS factors are given in EC7 for deformation parameters, therefore the deformation parameters should be factored as the strength or resistance parameters. An example where this type of analysis might be used is in the design of pile foundations. In this case the load deflection curve from the pile load test would be reduced by the factors discussed in Chapter 7 to obtain the load deflection curve for design.

6.6   Serviceability Limit State Design

6.6.1 Deformation Analyses

When using a direct method to design a foundation, C6.4(3)P requires that a separate deformation analysis is carried out to check against the occurrence of a serviceability limit state. The condition to be satisfied is that given in Chapter 2:

$$E_d \leq C_d \qquad (6.11)$$

where E_d is the foundation displacement calculated using unfactored loads and material parameters and C_d is the relevant serviceability criterion, for example the limiting values for movements given in §2.5.9.

Calculations for the settlement of foundations on saturated soils should consider the following three components:

- Undrained settlements (due to shear deformation with no volume change)
- Consolidation settlements
- Creep settlements.

A note of caution is given in C6.6(6) which points out that calculations of settlements should not be regarded as accurate but as merely providing an approximate indication.

Annex D provides a few sample methods for evaluating settlement. The first is the stress-strain method for calculating the total settlement of foundations on cohesive or non-cohesive (fine grained and coarse grained) soils. This is the preferred method and is based on computing the stresses at selected points in the ground beneath the foundation from elasticity theory, computing the strains in the ground from the stresses using stiffness moduli or other stress-strain relationships for the soil, and then integrating the strains to find the settlements.

Another method for calculating the total settlement of a foundation on cohesive or non-cohesive soil, referred to as the adjusted elasticity method, is to use an equation of the form:

$$s = \frac{p \, B \, f}{E_m} \qquad (6.12)$$

where:

p   is the SLS bearing pressure on the base of the foundation, reduced by the weight of excavated soil for normally consolidated soil; buoyancy effects should also be taken into account;

E_m is the drained Young's modulus of the soil estimated from the results back-analyses, or the results of laboratory or in situ tests;

f   is a coefficient whose value depends on the shape and dimensions of the foundation, the variation in stiffness with depth, the thickness of the compressible stratum, Poisson's ratio, the distribution of the bearing pressure and the point for which the settlement is calculated.

This adjusted elasticity equation should only be used if the stresses in the ground are such that no significant yielding occurs. It is not necessary for ground stress-strain behaviour to be linear, but the selected E_m value should be appropriate for the strain level and stress path. Great caution is advised in Annex D if the adjusted elasticity method is used for non-uniform ground conditions.

C6.6.1(7)P requires that differential settlements and relative rotations be assessed to ensure that they do not lead to the occurrence of a SLS. These may occur due to both a non uniform load distribution and the variability of the ground. According to C6.6.1(8) differential settlements of spread foundations may typically be up to 10mm, but not usually exceeding 50% of the calculated total settlement. Calculated differential settlements that do not take account of a structure's stiffness tend to be overpredictions. An analysis with ground-structure interaction can be used to justify reduced differential settlements and hence a more economic design.

According to C6.6.1(8) the tilting (= rotation) of a foundation may be estimated assuming a linear bearing pressure on the foundation base and calculating the settlements at the corner points using the vertical stress distribution in the ground beneath each corner point and the settlement methods given in **Annex D**. A equation by Poulos and Davis (1974) for calculating tilting, similar to the Adjusted Elasticity Method equation for calculating the settlement, is given in Example 6.2.

6.6.2 Indirect Design Methods

In traditional designs, spread foundations have often been designed using overall factors of safety, sometimes called mobilisation factors, in a bearing resistance calculation which are large enough to ensure that settlements are not excessive and hence to avoid the use of settlement analyses. This is an example of an indirect design method outlined in §2.5.8. EC7 provides no guidance on or requirements for the use of this method for the design of spread foundations and therefore it should only be used with caution. This method, together with certain limitation on its use, may be included in the EN version of EC7.

This method involves calculating, F the ratio of the characteristic bearing resistance, R_k to the characteristic, i.e. serviceability, loading, V_k:

$$F = R_k / V_k \qquad (6.13)$$

It is likely that the revised version of EC7 will include some guidance on the use of this method. For example, for conventional foundations on clay that is not very soft, if F is greater than 3, settlement calculations may not be required. This approach should only be used if well established comparable experience exists for this method in similar ground conditions and for similar structures.

6.7 Foundations on Rock

The additional factors which need to be considered when designing spread foundations on rock are given in **C6.7(1)P**. Most of these items are included as Item 4 in Table 2.4. Spread foundations may normally be designed using the method of presumed bearing pressures [**C6.7(2)**]. **Annex E** of EC7 gives a sample method for deriving the presumed bearing resistances for spread foundations on different types of rock based on the rock's uniaxial compressive strength and discontinuity spacing. In the case of strong intact rocks, the bearing pressure is limited by the compressive strength of the concrete foundation.

6.8 Structural Design

The structural design of a foundation involves determining the bending moment for design of the foundation height and reinforcement. This bending moment is calculated from the loads applied to the foundation and does not normally include the water pressures or the weight of the backfill on the foundation. As the Case B design load is larger than Case C load, the Case B load will normally give a larger bending moment and hence will control the structural design.

The foundation width is normally obtained from the Case C load, which is smaller than the Case B load. Since the Case B load is greater than the Case C load, the bearing pressure for structural design using the Case B load is greater than that for geotechnical design, but the bearing resistance is not exceeded as the bearing resistance is greater for Case B than for Case C.

According to **C6.8(2)**, a linear bearing pressure distribution may be assumed for the structural design of rigid foundations. The requirements for the structural design of spread foundations are given in EC2–3 (1998). According to **C(103)** of Informative Annex 107 of EC2-3, the critical section is located at a distance 0.15c inside the face of the supported column, where c is the column width. The requirements with regard to the increased concrete cover for the reinforcement in cast-in-situ spread foundations are given in **C4.1.3.3(108)** and **(109)** of EC2-3.

6.9 Design Examples

Example 6.1:   Pad Foundation, Vertical Load

The square pad foundation shown in Figure 6.8 supports a characteristic permanent vertical load of 800kN and variable vertical load of 300kN. The foundation is 0.7m thick, with a rising square column of side width 0.4m, and is constructed from concrete with unit weight of 24kN/m³. The foundation is founded at a depth of 2.0m in a uniform deposit of firm clay with $c_{uk} = 57kN/m^2$, $c'_k = 0$ and $\phi'_k = 24°$. The groundwater level is at a depth of 0.6m, $\gamma_w = 9.81kN/m^3$ and the unit weight of the clay is 18kN/m³ above and 20kN/m³ below the groundwater level. The width of the foundation with regard to the ULS and the bending moment for the reinforcement design are required.

ULS Design of the Foundation

Design Situations
ULS design of this foundation involves checking against bearing resistance failure for both undrained and drained conditions. No SLS check is considered.

Undrained Conditions
Choose a foundation width of 2.2m.

Design Action
For undrained conditions, the characteristic permanent load, V_k is equal the sum of the applied permanent vertical load, foundation weight and total weight of backfill above the foundation.

a) Supported permanent load   = 800.0kN
b) Weight of foundation slab   2.2 x 2.2 x 0.7 x 24   =   81.3kN
c) Weight of rising column   0.4 x 0.4 x 1.3 x 24   =   5.0kN
d) Weight of backfilled soil:
-     above water table   (2.2 x 2.2 - 0.4 x 0.4) x 0.6 x 18   =   50.5kN
-     below water table   (2.2 x 2.2 - 0.4 x 0.4) x 0.7 x 20   =   65.5kN

$$V_k = a) + b) + c) + d) = \underline{1002.3kN}$$

The design load is obtained by combining the permanent and variable loads using Equation 2.2 and the partial factors in Table 2.6:

Case B: V_B = 1.35*1002.3 + 1.5*300 = <u>1803.1kN</u>

Case C: V_C = 1.0*1002.3 + 1.3*300 = <u>1392.3kN</u>

Design Resistance
The design resistance for undrained conditions is obtained using Equation 6.4. As the foundation is square, $s_c = 1.2$, the other factors are all unity, and q = 0.6*18 + 1.4*20 = 38.8kN/m² so that:

Case B: R_B = 2.2*2.2*(57*5.14*1.2 +38.8) = <u>1889.4kN</u>

Case C: R_C = 2.2*2.2*((57/1.4)*5.14*1.2 +38.8) = <u>1403.2kN</u>

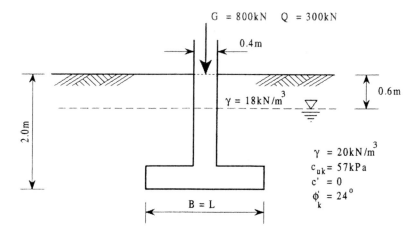

Figure 6.8: *Square pad foundation with vertical loads*

Conclusion

Since $V_B < R_B$ (1803.1 < 1889.4) and $V_C < R_C$ (1392.3 < 1403.2) both cases are satisfied. As V_C is closer to R_C than V_B is to R_B, Case C controls the size.

Drained Conditions

Design Action

1) Treating the effective weight of the backfill material and the water pressure forces on the top and bottom of the foundation separately, the components of the characteristic permanent load on the foundation are:

a) Supported permanent load		=	800.0kN
b) Weight of foundation slab	2.2 x 2.2 x 0.7 x 24	=	81.3kN
c) Weight of rising column	0.4 x 0.4 x 1.3 x 24	=	5.0kN

d) Effective weight of backfilled soil:

- above water table	(2.2 x 2.2 - 0.4 x 0.4) x 0.6 x 18	=	50.5kN
- below water table	(2.2 x 2.2 - 0.4 x 0.4) x 0.7 x (20 – 9.81)	=	33.4kN

e) Water pressure force on top of base

	(2.2 x 2.2 – 0.4 x 0.4) x 0.7 x 9.81	=	32.1kN
f) Uplift force on base	2.2 x 2.2 x 1.4 x 9.81	=	66.5kN

Characteristic permanent load, V_k = a) + b) + c) + d) + e) –f) = <u>935.8kN</u>

2) Alternatively, as the water pressure is hydrostatic, using the buoyant weights of the different materials below the groundwater level to calculate the components of the permanent load:

a) Supported permanent load		=	800.0kN

b) Buoyant weight of foundation slab

	2.2 x 2.2 x 0.7 x (24 – 9.81)	=	48.1kN

c) Design weight of rising column

- above water table	0.4 x 0.4 x 0.6 x 24	=	2.3kN
- below water table	0.4 x 0.4 x 0.7 x (24 – 9.81)	=	1.6kN

d) Design weight of backfilled soil
- above water table $(2.2 \times 2.2 - 0.4 \times 0.4) \times 0.6 \times 18$ $=$ 50.5kN
- below water table $(2.2 \times 2.2 - 0.4 \times 0.4) \times 0.7 \times (20 - 9.81)$ $=$ 33.4kN

$$V_k = a) + b) + c) + d) = \underline{935.9kN}$$

This result shows that the characteristic vertical load obtained using the alternative method is the same as that obtained when treating the effective weight of the backfill and the water pressure forces separately. The design loads for Cases B and C are obtained as above using the partial factors in Table 2.6:

Case B: $V_B = 1.35*935.9 + 1.5*300 = \underline{1713.5kN}$

Case C: $V_C = 1.0*935.9 + 1.3*300 = \underline{1325.9kN}$

Design Resistance

The design resistance for drained conditions is obtained using Equation 6.5 and the equations for the resistance and shape factors. As $c' = 0$, only the q and γ terms are relevant. Since $\phi'_k = 24°$, $N_{qB} = 9.60$, $N_{\gamma B} = 7.66$, $s_{qB} = 1.407$ and $s_{\gamma B} = 0.7$. For Case C, $\phi'_d = \tan^{-1} (\tan 24/1.25) = 19.6°$, which gives $N_{qC} = 6.15$, $N_{\gamma C} = 3.67$, $s_{qC} = 1.336$ and $s_{\gamma C} = 0.7$. The value of $q' = 0.6*18 + 1.4*(20 - 9.81) = 25.1kN/m^2$ so that:

Case B: $R_B = 2.2*2.2*(25.1*9.60*1.407 + 0.5*2.2*(20-9.81)*7.66*0.7)$
$= \underline{1929.9kN}$

Case C: $R_C = 2.2*2.2*(25.1*6.15*1.336 + 0.5*2.2*(20-9.81)*3.67*0.7)$
$= \underline{1136.5kN}$

Conclusion

$V_B < R_B$ (1713.5 < 1929.9), but $V_C > R_C$ (1325.9 > 1136.5), so the width, B = 2.2m that satisfies the undrained condition, with a partial factor of 1.4 on c_u, does not provide a sufficient level of safety for the drained condition. Therefore the width must be increasing to 2.4m, giving R_C = 1376.6kN, which is greater than the increased V_C = 1351.5kN, and so the ULS design is now OK. For Case B, V_B = 1748.1kN and R_B = 2328.2kN. Thus the design width is 2.4m. As for undrained conditions, Case C also controls the foundation size for drained conditions.

Structural Design

The bending moment for structural design of the pad foundation, M is the bending moment due to the Case B loads transmitted through the column at a cross section 0.15 times the column width inside the column face. The Case B loads are:

a)   Design supported permanent load $= 1.35*800 = 1080.0kN$

b)   Design supported variable load $= 1.5 * 300 = 450kN$

c)   Design weight of rising column
-    above water table $1.35*0.4 \times 0.4 \times 0.6 \times 24$ $=$ 3.1kN
-    below water table $1.35*0.4 \times 0.4 \times 0.7 \times (24 - 9.81)$ $=$ 2.1kN

Hence the design load for structural design is $V_B = a) + b) + c) = \underline{1535.2kN}$

$$M = \frac{V_B}{B^2} B\left(\frac{B-0.4}{2}+0.15*0.4\right)^2 = \frac{1535.2}{2.4^2} 2.4\left(\frac{2.4-0.4}{2}+0.15*0.4\right)^2$$

$$= 718.7 \text{kNm/m}$$

Thus the design moment for structural design of the foundation is 718.7kNm/m.

Example 6.2: Pad Foundation, Vertical and Horizontal Loads

The square pad foundation shown in Figure 6.9 supports a permanent vertical load of 500kN, a variable vertical load of 1500kN and an independent variable horizontal load of 120kN. The vertical loads are central on the foundation and the horizontal load is on the top of the foundation. The foundation is constructed from concrete with unit weight 24kN/m³ and founded at a depth of 1.0m in a uniform soil deposit with $c'_k = 10kPa$, $\phi'_k = 28°$ and $E_m = 28MPa$. The soil has a unit weight of 20kN/m³ and the groundwater table is at great depth. The limiting settlement and rotation of this foundation to prevent the occurrence of a SLS are 25mm and 1/2000 respectively.

ULS Design of the Foundation

Design Situation
Only drained conditions are relevant for this soil. As in the previous vertically loaded foundation example, the width with regard to the ULS in this eccentrically loaded example is found to be controlled by Case C, so the calculations are only shown for this case. A foundation width of 2.39m is chosen.

Design Loads
As the vertical and horizontal variable loads are independent, the combination factor $\psi = 0.7$ (EC1, Table 9.3) is used to obtain the design horizontal load.

Case C: $V_C = 1.0*(2.39*2.39*1.0*24 + 500) + 1.3*1500 = 2587.1\text{kN}$

$H_C = \psi*\gamma_Q*H_k = 0.7*1.3*120 = 109.2\text{kN}$

$e = H_Ch/V_C = 109.2*1.0/2587.1 = 0.042\text{m}$, and $e/B = 0.018$

$B' = B - 2*e = 2.39 - 2*0.042 = 2.306\text{m}$

$A' = B * B' = 2.39*2.306 = 5.511\text{m}^2$

Design resistance
The design resistance for drained conditions is obtained using Equation 6.5 and the equations for the various factors:

Case C: $\phi'_k = 28°$, $\phi'_C 23.0°$, so $N_{cC} = 18.10$, $N_{qC} = 8.70$, $N_{\gamma C} = 6.55$

$c'_C = 10/1.6 = 6.25\text{kPa}$

$s_q = 1 + (B'/B)\sin23° = 1 + (2.306/2.39)\sin23° = 1.378$

$s_\gamma = 1 - 0.3*(B'/B) = 1 - 0.3*(2.306/2.39) = 0.711$

$s_c = (s_qN_q - 1)/(N_q - 1) = (1.378*8.70 - 1)/(8.70 - 1) = 1.427$

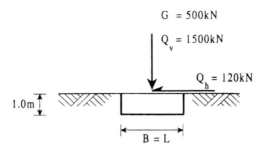

Figure 6.9: *Square pad foundation with a vertical and horizontal loads*

$i_q = (1 - H_C/V_C)^m = (1 - 109.2/2587.1)^{1.509} = 0.937$

$i_\gamma = (1 - H_C/V_C)^{m+1} = (1 - 109.2/2587.1)^{2.509} = 0.897$

$i_c = i_q - (1 - i_q)/N_c\tan\phi'_C = 0.937 - (1 - 0.937)/(18.14*\tan23°) = 0.929$

$R_C = A'(c'N_cs_ci_c + q'N_qs_qi_q + 0.5\gamma'B'N_\gamma s_\gamma i_\gamma)$

$\quad = 5.511*(6.25*18.10*1.427*0.929 + 1.0*20*8.70*1.378*0.937 +$
$\quad\quad 0.5*20*2.306*6.55*0.711*0.897)$

$\quad = 5.511*(150.0 + 224.7 + 96.3) = \underline{2595.7kN}$

Conclusion from ULS Design
As $V_C < R_C$ (2587.1 < 2595.7) and $V_B < R_B$ (3037.1 < 3055.5 – calculations not provided), Cases B and C are both satisfied. Case C determines the size of the foundation with regard to the ULS.

SLS Design of the Foundation

Design Situation
The SLS is checked by calculating the settlement of the foundation due to the average net bearing pressure from the characteristic permanent and variable loads, assuming that the foundation is rigid. The SLS criterion, C_{dS} is that the maximum settlement should not exceed 25mm and the rotation should not exceed 1/2000.

Design Actions
The SLS design load due to the supported characteristic permanent and variable loads and the net foundation weight is:

$$V_k = 500 + 1500 + 2.39*2.39*1.0*(24.0 - 20.0) = \underline{2022.9kN}$$

The design moment for the SLS calculation is obtained from the horizontal load:

$$M_k = 120*1.0 = \underline{120.0kNm}$$

Foundation settlement
The foundation settlement is calculated using the Adjusted Elasticity Method and Equation 6.12, with f = 0.86 for a square rigid foundation and $v = 0.3$:

$$s = \frac{pBf}{E_m} = \frac{V_k}{B^2}\frac{Bf}{E_m} = \frac{2022.9}{2.39^2}*\frac{2.39*0.86}{28000}*10^3 = 26.0mm$$

This settlement exceeds 25mm so the foundation width should be increased. If the width is increased to 2.50m, the settlement is 24.9mm and hence the SLS settlement condition is satisfied.

Foundation rotation
The foundation rotation is calculated using the following equation by Poulos and Davis (1974), which is similar to the Adjusted Elasticity Equation 6.12 used to calculate the settlement:

$$\theta = \frac{M_k \left(1 - v^2\right) I_\theta}{B^2 LE_m} \tag{6.14}$$

where M_k is the moment of the loads on the foundation; v is Poisson's ratio, I_θ is a geometrical factor to account for shape of the foundation and the thickness of the compressible layer, and B and L are the side lengths of the loaded area of the base, assuming a linear pressure distribution in the case of eccentric loading. For the loading in this example:

$$\theta = \frac{M_k \left(1 - v^2\right) I_\theta}{B^2 LE_m} = \frac{120 * \left(1 - 0.3^2\right) 1.0}{2.50^2 * 2.50 * 28000} = 2.496 * 10^{-4} \text{ radians} = 1/4006$$

As the rotation $\theta = 1/4006$ is less than 1/2000, the SLS rotation condition is satisfied.

Conclusion
In this example, the SLS condition controls the design. The SLS condition is satisfied if a <u>design width of 2.50m</u> is chosen.

Example 6.3: Pad Foundation with a Highly Eccentric Loading
The square pad foundation shown in Figure 6.10 supports a permanent vertical load of 300kN, a variable vertical load of 150kN and an independent variable horizontal load of 75kN at a height of 5m above the founding level. The foundation is constructed from concrete with a unit weight of $24kN/m^3$ and founded at a depth of 1.2m in a uniform soil deposit with $c' = 0$ and $\phi'_k = 34°$. The groundwater level is at the founding level and $\gamma = 19kN/m^3$ above and $21kN/m^3$ below the groundwater level. The unit weight of the water is $9.81kN/m^3$. The width of the foundation is required for drained conditions.

ULS Design of the Foundation

Design Situation
Only drained conditions are considered and a foundation width of 2.64m is chosen.

Design Action
When analysing the stability of this highly eccentrically loaded foundation, the vertical load is favourable so, for Case B, the appropriate partial factor on the permanent vertical load is 1.0. The variable vertical load is ignored when

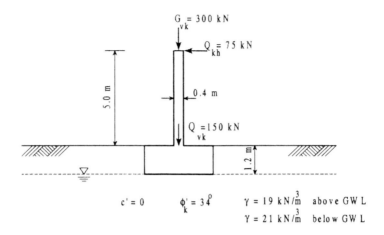

$G_{vk} = 300$ kN

$Q_{kh} = 75$ kN

5.0 m

0.4 m

$Q_{vk} = 150$ kN

1.2 m

$c' = 0$ $\phi'_k = 34°$ $\gamma = 19$ kN/m^3 above GWL

$\gamma = 21$ kN/m^3 below GWL

Figure 6.10: *Square pad foundation with a highly eccentric loading*

analysing the stability of the foundation, as this is the worst situation. Therefore the design loads, eccentricities, effective lengths and areas are:

Case B: $V_B = 300 + 2.64*2.64*1.2*24 = \underline{500.7kN}$

$H_B = 1.5*75 = \underline{112.5kN}$

$e_B = H_{dB}h/V_{dB} = 112.5*5/500.7 = 1.123$m, and $e_B/B = 0.426$

$B'_B = B - 2*e_B = 2.64 - 2*1.123 = 0.394$m

$A'_B = B * B' = 2.64*0.394 = 1.040$m^2

Case C: $V_C = V_B = \underline{500.7kN}$

$H_C = 1.3*75 = \underline{97.5kN}$

$e_C = H_{dC}h/V_{dC} = 97.5*5/500.7 = 0.974$m, and $e_C/B = 0.369$

$B'_C = B - 2*e_C = 2.64 - 2*0.974 = 0.692$m

$A'_C = B * B' = 2.64*0.692 = 1.827$m^2

The loading is highly eccentric because e/B exceeds 0.33 in each case.

Design Resistance

The design resistance for drained conditions is obtained using Equation 6.5 and the equations for the various factors:

Case B: $\phi'_B = \phi'_k = 34°$, so $N_{qB} = 29.4$, $N_{\gamma B} = 38.4$

$s_{qB} = 1 + (B'/B)\sin34° = 1 + (0.394/2.64)\sin34° = 1.083$

$s_{\gamma B} = 1 - 0.3*(B'/B) = 1 - 0.3*(0.394/2.64) = 0.955$

$m_B = [2+(B'/B)]/[1 + (B'/B)] = [2 + (0.394/2.64)]/[1 + (0.394/2.64)] = 1.870$

$i_{qB} = (1 - H_B/V_B)^{mB} = (1 - 112.5/500.7)^{1.870} = 0.621$

$i_{\gamma B} = (1 - H_B/V_B)^{mB + 1} = (1 - 112.5/500.7)^{2.870} = 0.482$

$R_B = A'(q'N_qs_qi_q + 0.5\gamma'B'N_\gamma s_\gamma i_\gamma)$

$= 1.040*(1.2*19*29.4*1.083*0.621+0.5*(21-9.81)*0.394*38.4*0.955*0.482)$

$= 1.040*(450.8 + 39.0) = \underline{509.4kN}$

Case C: $\phi'_C = 28.4°$, so $N_{qC} = 15.3$, $N_{\gamma C} = 15.4$

$s_{qC} = 1 + (B'/B)\sin28.4° = 1 + (0.692/2.64)\sin28.4° = 1.125$

$s_{\gamma C} = 1 - 0.3*(B'/B) = 1 - 0.3*(0.692/2.64) = 0.921$

$m_C = [2+(B'/B)]/[1 + (B'/B)] = [2 + (0.692/2.64)]/[1 + (0.692/2.64)] = 1.792$

$i_{qC} = (1 - H_C/V_C)^{mC} = (1 - 97.5/500.7)^{1.792} = 0.678$

$i_{\gamma C} = (1 - H_C/V_C)^{mC+1} = (1 - 97.5/500.7)^{2.792} = 0.546$

$R_C = A' (q'N_q s_q i_q + 0.5\gamma'B'N_\gamma s_\gamma i_\gamma)$

$= 1.827*(1.2*19*15.3*1.125*0.678+0.5*(21-9.81)*0.692*15.4*0.921*0.546)$

$= 1.827*(266.1 + 30.0) = \underline{541.0kN}$

Conclusion
As $V_B < R_B$ (500.7 < 509.4) and $V_C < R_C$ (500.7 < 541.0), Cases B and C are both satisfied. The calculations show that Case B controls the size of the foundation in this case. Also, as e_B/B is greater than 1/3, the side lengths of the foundation should be increased by 0.2m (§6.5.4). Hence the design width is 2.64 + 0.2 = 2.66m.

Example 6.4:   Cantilever Retaining Wall

The cast-in-situ concrete cantilever retaining wall shown in Figure 6.11 is 3m high, founded on soil with c' = 0 and $\phi'_k = 36°$, supports a granular backfill with c' = 0, $\phi'_k = 30°$ and $\gamma = 18kN/m^3$. There is a surcharge of 15kPa on the ground surface behind the wall. The wall is 0.3m thick and the base is 0.4m thick. The concrete unit weight is $24kN/m^3$. The width of the base of the wall is required for stability against bearing resistance failure and sliding.

Bearing Resistance

Design Situation
Only drained conditions are relevant for this example. The earth pressure is assumed to act horizontally on the virtual back of the wall at the heel. The most critical loading condition is when the surface variable loading extends just to the virtual back of the wall as shown in Figure 6.11. Calculations are only shown for Case C, as this case is found to control the width of the wall base. Calculations are carried out for a unit width of the wall – i.e. for a strip foundation. A width of 3.07m is chosen for the base of the retaining wall.

Design Action
Case C: W_s = weight of wall stem = $0.3*(3.0 - 0.4)*24 = 18.7kN/m$

W_b = weight of wall base = $0.4*3.07*24 = 29.5kN/m$

W_f = weight of fill = $(3.07 - 0.3)*(3.0 - 0.4)*18 = 129.6kN/m$

$V_C = 1.0*(W_s + W_b + W_f) = 18.7 + 29.5 + 129.6 = \underline{177.8kN/m}$

As $\phi'_k = 30°$ for fill, $\phi'_C = 24.8°$, hence $K_a = (1 - \sin\phi'_C)/(1 + \sin\phi'_C) = 0.409$

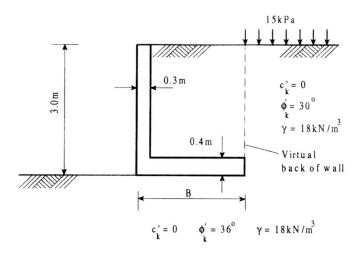

Figure 6.11: *Cantilever Retaining Wall*

H_e = design horizontal earth pressure force = $0.5*0.409*18*3.0^2$ = 33.1kN/m

H_s = design horizontal force due to surcharge = $1.3*(0.409*3.0*15)$ = 23.9kN/m

$$H_C = H_e + H_s = 33.1 + 23.9 = \underline{57.0kN/m}$$

The eccentricity, e of the load on the base is obtained from M_d/V_C where M_d is the design moment of the loads about the centre of the base:

$$e = (18.7*(3.07 - 0.3)/2) - 129.6*0.3/2 + 33.1*3.0/3 + 23.9*3.0/2)/177.8$$
$$= (25.9 - 19.4 + 33.1 + 35.9)/177.8 = 0.425m$$
$$B' = B - 2*e = 3.07 - 2*0.425 = 2.22m$$
$$A' = B' * 1.0 = 2.22 *1.0 = 2.22m^2/m$$

Design Resistance

Since e/B = 0.425/3.07 = 0.138 is less than 0.33, no special precaution needs to be taken with regard to the width of the base for bearing resistance. As the effective pressure due to the soil above the foundation level in front of the wall is ignored, q = 0 and hence the q component in the bearing resistance equation is zero. The c' component is also zero as c' = 0. Hence design resistance only has the γ component.

<u>Case C</u>: As $\phi'_k = 32°$, $\phi'_C = 26.6°$, so $N_{\gamma C} = 11.59$

For a strip foundation, $s_\gamma = 1.0$

m = 2

$i_\gamma = (1 - H_C/V_C)^{2+1} = (1 - 57.0/177.8)^3 = 0.314$

$R_C = A' (0.5\gamma'B'N_\gamma s_\gamma i_\gamma) = 2.22*0.5*20*2.22*11.59*1.0*0.314$

$= \underline{179.4kN/m}$

Conclusion

As $V_C < R_C$ (177.8 < 179.4), Case C is satisfied. When analysing this wall for bearing resistance for Case B, the vertical loads are unfactored, as they are favourable, so $V_B = V_C$. For $\phi'_B = 30°$, $K_{aB} = 0.333$, hence applying the factor of 1.35 to the earth pressure, $H_{eB} = 36.45$kN/m and 1.5 to the surcharge pressure, $H_{eB} = 36.45$kN/m, which are similar to the Case C values of 33.1kN/m and 23.9kN/m. These H and V values result in the eccentricity for Case B being similar to that for Case C. Although the loading for Cases B and C is similar, a smaller foundation size, only 2.6m, is required for Case B as the soil strength parameters are not factored. Hence Case C controls the design with regard to bearing resistance.

Sliding Resistance

Design Situation

Sliding resistance of the retaining wall is examined by ensuring that the design horizontal force does not exceed the design horizontal resistance.

Design Forces

Case C: $H_C = H_e + H_s = 33.1 + 23.9 = \underline{57.0\text{kN/m}}$

Design resistance

It is assumed that there is no passive earth pressure resistance in front of the wall so that $E_{pd} = 0$, and the horizontal resistance, R_C is only provided by the shear resistance, S_C between the base of the wall and the ground. For drained conditions, $S_C = V'_d \tan \delta_d$ and for cast-in-situ concrete $\delta_d = \phi'_c = 26.6°$ **[C6.5.3(7)P]**:

$$R_C = V'_d \tan \delta_d = 177.8 \tan 26.6 = 89.0\text{kN/m}$$

Conclusion

As $H_C < S_C$ (57.0 < 89.0), Case C is satisfied and therefore the design width of the base of the retaining wall is 3.07m.

Example 6.5:   Swimming Pool Subjected to Buoyancy

The swimming pool shown in Figure 6.12 is 25m long, 10m wide and 1.5m deep and has concrete walls 0.4m thick and a base slab 0.7m thick. The top of the pool is level with the ground surface and the ground water level is prevented, by a drainage system, from rising above a depth of 0.12m below the ground surface. The fill around pool has properties c' = 0 and $\phi'_k = 30°$ with $\gamma = 18$kN/m³ above the water table and 20kN/m³ below the water table. The unit weights of the concrete and water are 24 and 9.81kN/m³, respectively. The stability of the pool is to be checked with regard to buoyancy.

Design Situation

The critical design condition is when the pool is empty and needs to be checked for buoyancy due to the uplift force from the water pressure on the base of the pool. In this situation the resistance to buoyancy provided by the friction between the

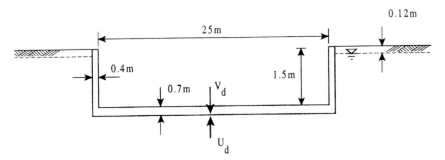

Figure 6.12: *Swimming pool with uplift and restoring forces*

backfill and the sides of the pool is small so that Case A is the critical design case. However, for completeness, Cases B and C are also examined.

Design Actions
With the water level at a depth of 0.12m and the pool founded at a depth of 2.2m, the characteristic uplift force due to the water pressure on the pool base is:

$$U_k = (2.2 - 0.12) *9.81*25.8*10.8 = \underline{5685.6kN}$$

The groundwater pressure is an unfavourable permanent action [**C2.4.2(16)**]. The weight of the concrete pool, a permanent favourable action, is:

$$W_p = (25.8*10.8*2.2 – 25*10*1.5)*24 = \underline{5712.2kN}$$

Using the partial action factors in Table 2.6, including the proposed value of 0.90 for permanent favourable actions rather than the value of 0.95 given in EC7 and treating upward forces as negative, the design actions for the three cases are:

Case A: $V_A = 0.9 * W_p - 1.0 * U_k = 0.9 * 5712.2 - 5685.6 = \underline{- 544.6kNkN}$

Case B: $V_{B1} = 1.0 * W_p - 1.35 * U_k = 5712.2 - 1.35*5685.6 = \underline{- 1963.4kN}$

$\qquad\;\; V_{B2} = 1.35 *(W_p - U_k) = 1.35*(5712.2 - 5685.6) = \underline{35.9kN}$

Case C: $V_C = 1.0 * W_p - 1.0 * U_k = 5712.2 - 5685.6 = \underline{26.6kN}$

Note that the partial factor value of 1.35 for Case B on the characteristic permanent unfavourable load, U_k leads to a U_d value corresponding to the water table at 0.61m above ground level and a design uplift force, V_{B1} of 1963.4kN. This uplift force is impossible, given that the water table is controlled so that it cannot rise above a depth of 0.12m below ground level. An alternative approach for Case B is that outlined in §2.5.6 and to treat the vertical forces as coming from a single source and to factor the net vertical force by 1.35. This gives a downward (positive) force V_{B2} of 35.9kN, not an uplift force, as the pool weight exceeds the water pressure force so that buoyancy does not occur for this situation and no soil resistance is required to prevent buoyancy. The design action for Case C, V_C is also positive as the design pool weight exceeds the uplift force and so no ground resistance is required to prevent buoyancy.

Design Resistance

It is assumed that the angle of shearing resistance between the fill and the pool, $\delta'_k = \phi'_k$ and $K_s = K_0 = 1 - \sin\phi'_k$. Hence, using the γ_m factors from Table 2.6, for Case A, $\delta'_A = 27.7°$ and $K_{0A} = 0.535$, for Case B, $\delta'_B = 30.0°$ and $K_{0B} = 0.5$, while for Case C, $\delta'_C = 24.8°$ and $K_{0C} = 0.581$. The total length of the pool sides is $2*10.8 + 2*25.8 = 73.2$m. Taking account of the water table at a depth of 0.12m, the design resistance due to friction between the backfill and the pool walls is then:

Case A: R_A = $AK_{0A}\sigma'_v\tan \delta'_A$

= $0.12*73.2*0.535*18*0.6*\tan27.7 +$

$2.08*73.2*0.535*0.5*(18*0.12 + (20 - 9.81)*2.08)*\tan27.7$

R_A = $2.7 + 545.7$ = <u>548.4kN</u>

Case B: R_B = $2.7 + 560.7$ = <u>563.5kN</u>

Case C: R_C = $2.5 + 521.0$ = <u>523.5kN</u>

Conclusion

As $V_A < R_A$ (544.6 < 548.4), this design is satisfactory with regard to buoyancy. As V_B and V_C are both downward forces, no buoyancy arises with Case C. The pool stability with regard to buoyancy for Case A depends on the water table being at a depth not less than 0.12m below ground level. If the resistance on the pool walls is ignored, the water table must be kept at least 0.22m below ground level. This result shows that the soil resistance has a minor effect on the pool stability with regard to buoyancy. This example also demonstrates the importance of checking for Case A in the case of structures, such as swimming pools, deep basements and underground structures, subject to hydraulic uplift forces.

References

BRE (1987) *The influence of trees on house foundations in clay soils*, Building Research Establishment Digest 298, BRS, Watford.

DIN 4017 (1979a) *Part 1: Shear failure calculations for shallow foundations with vertical and central loading*, DIN, Berlin.

DIN 4017 (1979b) *Part 2: Shear failure calculations for shallow foundations with oblique and eccentric loading*, DIN, Berlin.

ENV 1992-3 (1998) *Eurocode 2: Design of concrete structures – Part 3: Concrete foundations*, CEN, Brussels.

Poulos H.G & Davis E.H. (1974) *Elastic solutions for soil and rock mechanics*, Wiley.

Chapter 7

Pile Foundations

7.1 Introduction

Section 7 of EC7 provides the requirements for the design of all types of piles, including compression and tension piles, laterally loaded piles and piles installed by driving, jacking, screwing or boring, with or without grouting [**C7.1(1)P**]. Pile design from the results of static load tests, from empirical or analytical calculation methods and from dynamic load tests, is covered.

Section 7 lists the items to be considered when choosing a pile type, provides recommendations concerning the design of individual piles and pile groups, and the requirements for the supervision of pile construction. Special consideration is given to the treatment of ground displacements as actions, as this is an important aspect of pile design. Ground displacements relative to a pile induce forces in the pile, and these have to be resisted by the pile and by the ground.

The approach adopted in EC7 for the design of piles differs from that used for the design of spread foundations in that it is a resistance factor approach (§2.5.6) with partial factors applied to the characteristic pile resistance rather than to the characteristic ground strength parameters. The partial factors given in Section 7 enable the design pile resistance to be obtained from the characteristic resistance for Case C. Although not clearly stated in Section 7, pile designs should also be checked for Case B and, in buoyancy situations, for Case A.

A unique aspect of pile design to EC7 is that a method is given for determining the characteristic pile resistance directly from the results of pile load tests. This method takes account of the number of pile tests.

The requirements for the construction of piles are covered by the following execution standards, which are at the pre-standard stage:

- prEN 1536 (1997) Execution of special geotechnical works - Bored piles;
- prEN 12699 (1998) Execution of special geotechnical works - Displacement piles;
- prEN 288008 Execution of special geotechnical works – Micro-piles;
- prEN 12794– Precast concrete products - Precast concrete foundation piles.

Limit state	Considered
Loss of overall stability	
Bearing resistance failure of the pile foundation	
Uplift or insufficient tensile resistance of the pile foundation	
Failure of the ground due to transverse loading of the pile foundation	
Structural failure of the pile in compression, tension, bending, buckling or shear	
Combined failure in the ground and in the pile foundation	
Combined failure in the ground and in the structure	
Excessive settlements	
Excessive heave	
Unacceptable vibrations	

Table 7.1: *Limit states to be considered in pile design*

7.2   Limit States for Pile Foundations

The limit states listed in **C7.2.1(1)P** that need be considered in the design of pile foundations are presented in Table 7.1. Some of these limit states are not unique to pile foundations, such as loss of overall stability, the requirements for which are provided in Section 9 of EC7.

7.3   Actions and Design Situations

The design values of the actions on a pile due to the loads from the supported structure are determined, as for a spread foundation, using the partial load factors given in Table 2.6. Special consideration needs to be given to the actions on a pile due to relative ground displacements which cause compressive or tensile forces on a pile and/or bending of a pile. These ground displacements could arise from consolidation, swelling, adjacent loads, creeping soil, landslides or earthquakes. The actions resulting from these ground displacements include downdrag (the downward force due to ground settling relative to a pile, which is traditionally termed negative skin friction), heave and transverse forces. For these actions, **C7.3.2.1(1)P** requires that the design strength and stiffness values of the moving ground be upper values; i.e. when calculating downdrag, the selected characteristic soil strength should be a cautious estimate of the highest value likely to affect the pile.

The actions arising from ground displacement are to be included in pile design using one of the following approaches given in **C7.3.2.1(2)P**:

a) treating the ground displacement as an action and then using an interaction analysis to determine the forces, displacements and strains in the pile;
b) evaluating an upper bound to the force which the ground could transmit to the pile and using this as the design action.

If approach (b) is used and the downdrag force is treated as an action, then its value shall be the maximum which could be generated by large settlement of the ground relative to the pile [**C7.3.2.2(1)P**]. Where the ground settlement is small, a more economic design may be obtained by adopting approach (a) and carrying out an interaction analysis [**C7.3.2.2.(3)P**]. When analysing heave, the movement of the ground shall generally be treated as an action [**C7.3.2.3(1)P**]. The loads exerted on piles by transverse ground movements should normally be evaluated by considering the piles as beams in a deforming soil mass [**C7.3.2.4(2)**].

7.4   Design Methods and Design Considerations

C7.4.1(1)P requires that the design of piles shall be based on one of the following approaches:
- the results of static load tests which have been shown by calculations or otherwise to be consistent with other relevant experience;
- empirical or analytical calculations which have been validated by static load tests in comparable situations;
- dynamic load tests which have been validated by static load tests in comparable situations.

In view of the limitations associated with all empirical and analytical pile calculations and with dynamic load tests, EC7 emphasises the important role of static load tests in the design of piles by requiring that these methods be validated by static load tests. The requirements for static load tests are discussed in §7.8.2.

The use of the observed performance of existing pile foundations in place of load tests is sometimes acceptable, provided this is backed up by a geotechnical investigation and ground testing [**C7.4.1(2)**]. GC1 structures may be designed from comparable experience, without supporting load tests or calculations, provided the pile type, method of installation and ground conditions are within the area of experience of the designer, and provided the ground conditions are checked and supervised [**C7.4.1(2)**]. The various aspects which need to be considered in the design of piles are listed in the clauses in **C7.4.2**. These items are self-explanatory.

7.5   Design of Piles in Compression

7.5.1 Limit State Design
When designing a compression pile to EC7, **C7.6.1(1)P** requires that the following limit states are demonstrated to be sufficiently improbable:
- ultimate limit state of overall stability failure;
- ultimate limit state of bearing resistance failure of the piled foundation;
- ultimate limit state of collapse or severe damage to a support structure caused by displacement of the piled foundation;
- serviceability limit state in the supporting structure caused by displacement of the piles.

The design of pile foundations against overall stability failure is similar to the design of spread foundations against this failure mechanism except that failure modes involving slip planes passing both below the piles and through the piles should be considered. The requirements for design against overall failure are covered by **C6.5.1**.

Where an ultimate limit state occurs in the supported structure due to displacement of the pile foundation before the displacement required to mobilise fully its bearing resistance, the ULS partial factors should be applied to the whole load-settlement curve [**C7.6.1(2)**].

Serviceability limit state calculations are carried out as outlined in §2.5.7 with partial factors of unity on the actions and resistances.

7.5.2 Bearing Resistance Failure

The fundamental relationship to the satisfied with regard to the bearing resistance of a pile, given in **C7.6.3.1(1)P**, is:

$$F_{cd} \leq R_{cd} \tag{7.1}$$

where:

F_{cd} is the ULS design axial compression load; and

R_{cd} is ULS design bearing resistance and is the sum of all the bearing resistance components against axial loads, taking into account the effect of any inclined or eccentric loads.

F_{cd} is determined, using the partial factors in Table 2.6 for Cases B and C, from the characteristic applied loads and the self weight of the pile and includes any downdrag, heave or transverse loading on the pile, as appropriate. Example 7.1 provides an example of the determination of F_{cd}.

It is common practice in the design of compression piles to assume that the weight of the pile cancels the weight of the overburden at the foundation base and to exclude these from F_{cd} and R_{cd} respectively. This is permitted by EC7, provided the two cancel approximately [**C7.6.3.1(2)**]. The implication of this is to reduce the effect of the pile weight in the design, as shown in Example 7.3. According to **C7.6.3.1(2)**, the pile weight may not cancel the weight of the overburden if:

- downdrag is significant (as it reduces the effective stress at the base of a pile);
- the soil is light, such as a pile through peat; or
- the pile extends above the surface of the ground.

R_{cd} is obtained from load tests, from soil parameters or from dynamic load tests for each of the design cases.

Pile groups

For piles in groups, **C7.6.3.1(3)P** requires that bearing resistance failure of the piles failing individually and acting as a block shall both be checked and the lower of these be taken as the design bearing resistance. Generally the bearing resistance of a pile group acting as a block may be analysed as a large pile [**C7.6.3.1(4)**]. When piles are used to reduce the settlement of a raft, their resistance corresponding to the creep load may be used in analysing the serviceability states of the structure [**C7.6.3.1(4)**]. The creep load is defined in the ISSMFE recommendations on pile testing (ASTM, 1985) as the critical experimental load beyond which the rate of settlement under constant load takes place at a notably increased increment.

The need to consider the effect on the bearing resistance of adjacent piles and weak layers under the base is highlighted in **C7.6.3.1(5)P** & **C7.6.3.1(6)P**. The nature of the structure must also be considered when designing pile groups [**C7.6.3.1(7)P**]. For flexible structures, the weakest pile governs the limit state, while for stiff structures, a failure mode involving one pile need not be considered as the load can be redistributed. Special attention needs to be given to failure of edge piles by inclined or eccentric load.

Example 7.1: Design Load for a Pile Subjected to Downdrag

The ULS design axial compressive load is required for a 250mm square precast concrete pile, $\gamma = 23.4 kN/m^3$, that is to support a vertical permanent load of 350kN and a variable vertical load of 150kN. The pile is to be driven through 4m of loose sand fill which is to be placed on 10m of very soft organic clay overlying strong limestone rock, as shown in Figure 7.1. The sand properties are $\gamma = 18kN/m^3$, $c'_k = 0$, $\phi'_k = 33°$ and the clay properties are $\gamma = 15kN/m^3$, $c_{uk} = 8kPa$ for vertical effective pressures, σ'_v less than the preconsolidation pressure and $c_{uk}/\sigma'_v = 0.3$ for σ'_v greater than the preconsolidation pressure. The water table is 1m below the original ground level. Large settlements are expected in the soft clay layer.

Design Situation
The above example has been selected for its simplicity as the pile is founded on rock, which can be treated as a rigid base. As the pile rests on a rigid base, the upper bound to the downdrag which can be transmitted from the ground to the pile is used. This would not be the case if, for example, the pile base was on stiff clay as the pile itself would undergo displacement as the downdrag was mobilised, thus reducing the magnitude of the downdrag taken by the pile and mobilising some positive pile shaft resistance on the lower part of the pile. Such an example may require an interaction analysis to determine the downdrag.

Downdrag
The downdrag is given by $\Sigma q_s A_i$ where q_s is the resistance per unit area of the pile shaft and A_i is the pile surface area in layer i. Note that q_s is used in EC7 for the shaft resistance instead of the traditional symbol, τ. For the sand, drained conditions apply and $q_s = c'_s + K_s \sigma_v' \tan \delta'$ where:

c'_s is the cohesion between the pile shaft and the soil;
K_s is the lateral earth pressure coefficient on the pile shaft;
σ_v' is the vertical effective stress; and
δ' is the angle of shearing resistance between the pile shaft and the soil

For the soft clay, undrained conditions apply and $q_s = c_a$ where the c_a is the adhesion between the pile shaft and the clay.

It should be noted that the characteristic downdrag (D_k) is obtained using characteristic strength parameter values that are cautious estimates of the highest values which could affect the stability of the pile. In this example it is assumed that $c'_{sk} = c'_k$, $\delta'_k = \phi'_k$ and $c_{ak} = c_{uk}$. It is also assumed that the σ_v' value in the clay after

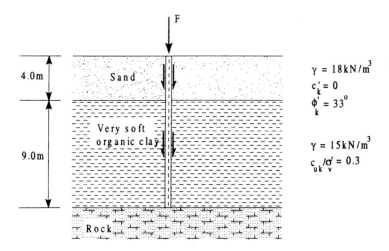

Figure 7.1: *Pile subjected to downdrag*

placing the sand fill exceeds the preconsolidation pressure, so that $c_{uk} = 0.3\sigma_v'$.

Downdrag from the sand layer
The average σ_v' value in the middle of the 4m thick sand layer is $2 \times 18 = 36$kPa. A_i = A_1 = $4*0.25*4.0 = 4.0m^2$. As ϕ_k' = 33 °, K_{sk} = $(1 - \sin\phi_k') = (1 - \sin 33°) = 0.455$ and $\tan\delta' = \tan 33° = 0.649$. Hence the downdrag from the sand layer is:

$$D_{k1} = q_{sk}A_1 = (K_s\sigma_v'\tan\delta')A_1 = 0.455*36*0.649*4*0.25*4.0 = \underline{42.5kN}$$

Downdrag from the clay layer
The average σ_v' value in the middle of the 10m thick clay layer is $4 \times 18 + 5*15 -$ $4*9.81 = 107.8$kPa. $A_i = A_2 = 4*0.25*10.0 = 10.0m^2$. Hence the downdrag due to the clay is:

$$D_{k2} = q_{sk}A_2 = c_{uk}A_2 = \Sigma 0.3\sigma_v'A_i = 0.3*107.8*4*0.25*10.0 = \underline{323.4kN}$$

Pile weight
Weight of the pile: $W_p = 0.25*0.25*14*23.4 = \underline{20.5kN}$

Design Compressive Load
Case B: $F_{cB} = 1.35*(350 + 42.5 + 323.4 + 20.5) + 1.5*150 = \underline{1219.1kN}$
Case C: $F_{cC} = 1.0*(350 + 42.5 + 323.4 + 20.5) + 1.3*150 = \underline{931.4kN}$

7.5.3 Ultimate Bearing Resistance from Pile Load Tests
The requirements in EC7 for pile load tests are discussed in §7.8. The factors that need to be considered when using pile test results to design a pile are listed in **C7.6.3.2**. One of these factors is that, for a pile which will experience downdrag, the shaft resistance mobilised in and above the compressible stratum during the load tests will not be available in the design situation and hence shall be deducted from

Number of pile load tests	1	2	>2
(a) Factor ξ on mean R_{cm}	1.5	1.35	1.3
(b) Factor ξ on lowest R_{cm}	1.5	1.25	1.1

Table 7.2: *Values of the correlation factor ξ to obtain the characteristic bearing resistance*

Partial resistance factor	γ_b	γ_s	γ_t
Driven piles	1.3	1.3	1.3
Bored piles	1.6	1.3	1.5
CFA (Continuous flight auger)	1.45	1.3	1.4

Table 7.3: *Values of the partial resistance factors for different types of piles*

the recorded pile resistance. **C7.6.3.2(4)P** requires that, in such cases, the measured shaft resistance or the most unfavourable estimate of the design positive shaft resistance in the compressible stratum be subtracted from the forces measured at the pile head. It is recommended that the maximum load applied to a working test pile should be in excess of the design external load plus twice the downdrag [**C7.6.3.2(5)**].

The following stages are involved in the design of piles to EC7 from load tests:

1) Select pile type, design load tests, carry out load tests;
2) Determine the measured pile resistance, R_{cm} from the load tests;
3) Determine the characteristic pile resistance using the correlation (reduction) factors, ξ in Table 7.2, $R_{ck} = R_{cm}/\xi$;
4) Separate R_{ck} into base and shaft components, R_{bk} and R_{ck}, where possible;
5) The design bearing resistance for Case B, $R_{cB} = R_{ck}$, while the design bearing resistance for Case C, R_{cC} is obtained using the partial factors in Table 7.3 for the base, shaft and total resistance, γ_b, γ_s, γ_t:

$$R_{cC} = R_{bk}/\gamma_b + R_{sk}/\gamma_s \quad \text{or} \quad R_{ck}/\gamma_t \qquad (7.2)$$

R_{cm}, the total pile bearing resistance measured in a pile test, is reduced to account for the reduction in shaft resistance where downdrag is involved, or adjusted to take account of other factors, such as differences in the ground at different parts of the site and differences due to variations in the installation procedure. Allowance should also be made to the R_{cm} value for size difference where the trial pile is smaller in diameter than the working pile.

The factor ξ used to determine R_{ck} from the R_{cm} values, accounts for the uncertainty in determining bearing resistance for the working piles from the results of a limited number of load tests on similar piles installed in similar ground at a finite number of locations on the site. The uncertainty depends on the number of test results available, so ξ reduces as the number of test results increases. R_{ck} is the minimum value obtained applying the ξ values to the mean and average of the load test results. Schuppener (1998) has pointed out that in most cases when using the values in Table 7.2, R_{ck} will be based on the lowest R_{cm} value and this is excessively conservative where a structure can redistribute the load from weak to strong piles. These ξ values are likely to be revised when EC7 is issued as an EN.

According to **C7.6.3.2(8)P** the characteristic resistance, R_{ck} should be separated into base and shaft components, R_{bk} and R_{bk}, and separate partial factors, γ_b and γ_s from Table 7.3, are applied to these components. This is straightforward where the base or shaft resistances are measured separately. If these are not measured separately, **C7.6.3.2(9)** recommends that the ratio of the two be estimated using the ground test results. Normally, however, pile load tests only provide the pile load against the settlement without distinguishing between the base and shaft resistance so **C7.6.3.2(11)** allows piles to be designed by applying the γ_t in Table 7.3 to the total resistance, R_{ck}.

It should be noted that pile load tests in cohesive soils are normally carried out at a rate that is too fast to allow dissipation of the excess pore pressures generated during the test, hence such tests only provide information about a pile's undrained bearing resistance.

Example 7.2: Pile Design using Pile Load Tests

It is required to verify the design of the 600mm diameter CFA pile, 11m long, shown in Figure 7.2, to support a characteristic permanent vertical load of 1200kN and a characteristic variable vertical load of 200kN. Two load tests on similar trial piles close by were both taken to 4000kN without reaching failure. This load represents the maximum load which could be applied using the loading frame on site for the tests. The ground comprised 2m of loose and variable fill over a very stiff glacial till (boulder clay).

Design Loads
The weight of the pile is not included in the design load when using load test results because the pile resistance obtained by applying a load to the top of a similar pile in the load test takes account of the pile weight. The design loads are therefore:

Case B: $F_{cB} = 1.35 \times 1200 + 1.5 \times 200 = \underline{1920kN}$
Case C: $F_{cC} = 1.00 \times 1200 + 1.3 \times 200 = \underline{1460kN}$

Design Resistances
R_{cm} is 4000kN for both tests, therefore the mean and the lowest R_{cm} values are the same. Hence R_{ck} is the least of 4000/1.35 (= 2963kN) and 4000/1.25 (= 3200kN) so that R_{ck} =2963kN. Since instrumentation on one pile showed that only 600kN was taken base resistance at the maximum test load, this result is used to separate R_{ck} into R_{bk} and R_{ck} as follows:

$R_{bk} = 600/4000(R_{ck}) = 0.15 \times 2963 = \underline{444kN}$
$R_{sk} = (4000 - 600)/400(R_{ck}) = 0.85 \times 2963 = \underline{2519kN}$

The design resistances are then:

Case B: $R_{cB} = R_{ck} = \underline{2963kN}$
Case C: $R_{cC} = R_{bk}/1.45 + R_{sk}/1.3 = 444/1.45 + 2519/1.3 = \underline{2244kN}$
            or, using the total resistance, $= 2963/1.4 = \underline{2116kN}$

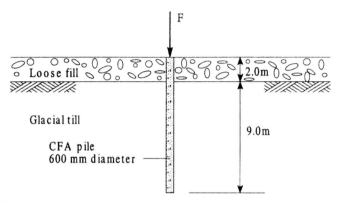

Figure 7.2: *CFA pile in glacial till*

These results show that the design resistance obtained for Case C using the total resistance is slightly less (i.e. more cautious) than that obtained when the bearing resistance is divided into its base and shaft components. The design resistance for Case B is not affected as the characteristic resistance values are not factored.

Design Check
The design is checked by comparing the design loads and resistances:

<u>Case B</u>: $F_{cB} \leq R_{cB}$ (1920kN < 2963kN), therefore Case B satisfied;

<u>Case C</u>: $F_{cB} \leq R_{cB}$ (1460kN < 2116kN), therefore Case C satisfied.

Conclusions
The ratio of $R_{cB}/F_{cB} = 1.54$, whereas $R_{cC}/F_{cC} = 1.45$, thus Case C controls the design of this pile. This is consistent with the discussion in §7.5.4 concerning the FOS.

The CFA piles in this example have an ultimate bearing resistance in excess of the maximum load applied in the tests, which was only 4000kN. According to **C7.5.2.1(1)P**, the loading in pile load tests shall be such that conclusions can be drawn about the ultimate failure load. **C7.5.2.1(2)** indicates that the failure load may be determined by extrapolation from the results of compression load tests which have not been taken to failure. However no guidance is given on the extent to which this extrapolation can be made.

7.5.4 Ultimate Bearing Resistance from Ground Test Results
The method for determining the ultimate bearing resistance from the results of field and laboratory tests involves the following stages:

1) Carrying out a field and/or laboratory test and determining derived parameters;
2) Interpreting the characteristic ground parameter values;
3) Selecting the pile type;
4) Calculating the base resistance per unit area, q_{bcal} and the shaft resistance per unit area for layer i, q_{sical} using the characteristic ground parameter values;
5) Obtaining the characteristic base resistance per unit area, q_{bk} and shaft resistance per unit area, q_{sik} by dividing the calculated values by 1.5:

$$q_{bk} = q_{bcal}/1.5 \quad \text{and} \quad q_{sik} = q_{sical}/1.5$$

6) Calculating the characteristic shaft and base resistances from:

$$R_{bk} = A_b q_{bk} \text{ and } R_{sk} = \Sigma q_{sik} A_{si}$$

where:

A_b = the nominal plan area of the base of the pile;

A_{si} = the nominal surface area of the pile in soil layer i.

7) Obtaining the design base and shaft resistances, R_{bd} and R_{sd} from R_{bk} and R_{sk} in the same manner as the design resistances from load test results, i.e.

<u>Case B</u>: $R_{cB} = R_{bk} + R_{sk} \text{ or } = R_{ck}$.

<u>Case C</u>: $R_{cd} = R_{bk}/\gamma_b + R_{sk}/\gamma_s \text{ or } = R_{ck}/\gamma_t$.

where the γ_b, γ_s, and γ_t values are those given in Table 7.3.

The characteristic ground parameter values should generally be evaluated in accordance with the process outlined in §3.3, but the results of load tests may also be taken into account in the evaluation process [**C7.4.1(2)**]. The special feature introduced in EC7 for determining R_{ck} from ground parameter values is Step (5) which requires the resistances calculated from characteristic ground parameters to be divided by a correlation factor of 1.5 to obtain R_{ck} or R_{bk} and R_{sk} [**C7.6.3.3(4)**].

The overall factor of safety (FOS), as defined by Equation 6.6, is obtained for Case B by multiplying the partial resistance factors in Table 7.2 by the partial load factors while, for Case C, this value is further multiplied by the correlation factor of 1.5. The FOS values obtained depend on the proportion of the load which is variable and on the pile type. The FOS values obtained by Frank (1997) for 0%, 33% and 50% variable load, based on γ_t (i.e. without separating the base and shaft resistances), are given in Table 7.4. These FOS values apply for both undrained and drained conditions.

The larger of the Case B and C FOS values in Table 7.4 for each loading condition and pile type is the critical value. Examination of the FOS values in Table 7.4 shows that Case C will always be critical, except for driven piles with a low percentage of variable load. The critical FOS values in Table 7.4 range from 2.03 to 2.59 and hence are similar to the traditional overall factors of safety varying from 2.0 to 2.5 which are used for pile design in the various national standards.

The FOS values for a pile designed on the basis of the bearing resistance obtained from a single pile load tests are the same as the values given in Table 7.4, as the ξ value for one test is 1.5. However, as the number of pile tests increases, the required FOS reduces, indicating the advantage of carrying out more than one test.

The importance of taking account of comparable experience when using calculation rules based on ground properties to design of piles is emphasised in **C7.6.3.3(5)P**. Various items which should be considered when calculating pile resistance using ground properties are listed in **C7.6.3.3(6)**. The significance of the strength of the zone of ground around the base of the pile on the ultimate bearing resistance is highlighted in **C7.6.3.3(7)P**. As an application rule to this requirement, **C7.6.3.3(8)** recommends that, if weak ground is present at a depth of less than 4 times the base diameter below the base of the pile, a punching failure mechanism should be considered. **C7.6.3.3(9)P** requires that the base resistance for open ended driven piles be the smaller of the shearing resistance between the ground plug and the inside face of the pile, or the base resistance calculated using the cross sectional area of the base, unless special devices are introduced to induce plugging. Also, in

	Equivalent Overall Factors of Safety (FOS)					
% Variable load	0%		33%		50%	
Design case	Case B	Case C	Case B	Case C	Case B	Case C
Types of pile						
Driven piles	2.03	1.95	2.10	2.15	2.14	2.24
Bored piles	2.03	2.25	2.10	2.48	2.14	2.59
CFA piles (continuous flight auger)*	2.03	2.10	2.10	2.31	2.14	2.42

* not in Frank (1997)

Table 7.4: *FOS values for different types of pile using partial factors in EC7*

the case of driven piles with oversized bases, the effect of driving this enlarged base on the shaft skin friction needs to be considered [**C7.6.3.3(10)P**].

Example 7.3: Pile Design using Ground Properties

The piles and ground conditions for this example are the same as those for Example 7.2, where the pile design is based on load tests. It is required in this example to verify that, on the basis of the ground properties, 600mm diameter CFA piles will support the characteristic permanent and variable vertical loads of 1200kN and 200kN, respectively. The c_{uk} value for the very stiff glacial till increases linearly from 100kPa at 2m depth (top of the till) to 600kPa at 11m (bottom of the pile). The properties of the glacial till are $\gamma = 22kN/m^3$, $c' = 0$, $\phi'_{ks} = 36°$ for shaft resistance and $\phi'_{kb} = 34°$ for the base resistance (the reduction in ϕ' is to allow for the higher stress levels at the base). The unit weights of the fill and concrete are 18kN/m³ and 24kN/m³, respectively. The water table is at a depth of 2.0m, which coincides with the top of the till.

Design Situation
The FOS values in Table 7.4 show that Case C will be critical, as this is not a driven pile. It is therefore not strictly necessary to check the pile design for Case B. Nevertheless, for completeness, the calculations for Case B are also presented.

Undrained Conditions

Design Actions
Unlike in Example 7.2, the pile weight, $W_p = \pi*0.6^2*11.0*24.0/4 = 74.6kN$, should be included an action in this example, so that the design loads are:

Case B: $F_{cB} = 1.35(1200 + 74.6) + 1.5*200 = \underline{2020.7kN}$

Case C: $F_{cC} = 1.0(1200 + 74.6) + 1.3 \times 200 = \underline{1534.6kN}$

Design Resistances
For undrained conditions it is assumed that the base resistance per unit area, q_b is $9c_u + \sigma_{v0}$, where σ_{v0} is the initial vertical total overburden pressure at the level of the pile base, and the shaft resistance per unit area, q_s is αc_u, where $\alpha = 0.4$ for this stiff clay. Hence:

$q_{bk} = q_{bcal}/1.5 = (9c_u + \sigma_v)/1.5 = (9*600 + 2*18 + 9*22)/1.5 = 3756.0\text{kPa}$

$q_{sk} = q_{scal}/1.5 = \alpha c_u/1.5 = 0.4*\{(100 + 600)/2\}/1.5 = 93.3\text{kPa}$

$$R_{bk} = A_b q_{bk} = (\pi*0.6^2/4)*3756.0 = 1062.0\text{kN}$$

$$R_{sk} = \sum A_s q_{sk} = \pi*0.6*9*93.3 = 1582.8\text{kN}$$

$R_{cB} = R_{bk} + R_{sk} = 1062.0 + 1582.8 = \underline{2644.8\text{kN}}$

$R_{cC} = R_{bk}/\gamma_b + R_{sk}/\gamma_b = 1062.0/1.45 + 1582.8/1.3 = \underline{1950.0\text{kN}}$

Design Check

As $F_{cB} < R_{cB}$ (2020.7 < 2644.8) and $F_{cC} < R_{cC}$ (1534.6 < 1950.0), both Cases B and C are satisfied.

<u>Drained Conditions</u>

Design Actions

For drained conditions the action due to the uplift water pressure force on the pile base should be included in the design load. This force is subtracted from the pile weight and applied loads. The characteristic water pressure force is $U_k = A_b*\gamma_w*$height of water table above pile base $= (\pi*0.6^2/4)*9.81*9.0 = 25.0\text{kN}$. Hence the design loads are:

<u>Case B</u>: $F_{cB} = 1.35(1200 + 74.6 - 25.0) + 1.5*200 = \underline{1987.0\text{kN}}$

<u>Case C</u>: $F_{cC} = 1.0(1200 + 74.6 - 25.0) + 1.3\text{x}200 = \underline{1509.6\text{kN}}$

Design Resistances

For drained conditions, the base resistance per unit area is $q_b = q'N_q$ where q' is the effective overburden pressure at the level of the pile base and $N_q = 40$ for $\phi'_k = 34°$ (Berezantsev et al., 1961). The shaft resistance per unit area of the shaft is assumed to be given by $q_s = K_s\sigma_{vo}'\tan\delta$. The shaft resistance in the fill is ignored. For the CFA pile in this overconsolidated clay, $K_s = 1.5$ and $\delta' = \phi'_k = 36°$. Hence:

$q_{bk} = q_{bcal}/1.5 = q'N_q/1.5 = (2*18.0 + 9*(22.0 - 9.81))*40/1.5 = 3885.6\text{kPa}$

$q_{sk} = q_{scal}/1.5 = K_s\sigma_{vo}'\tan\delta/1.5$

$\quad = 1.5*\{(2*18 + 2*18 + 9*(22 - 9.81))/2\}*(\tan 36)/1.5 = 66.0\text{kPa}$

$$R_{bk} = A_b q_{bk} = (\pi*0.6^2/4)*3885.6 = 1098.6\text{kN}$$

$$R_{sk} = \Sigma A_s q_{sk} = \pi*0.6*9*66.0 = 1119.7\text{kN}$$

$R_{cB} = R_{bk} + R_{sk} = 1098.6 + 1119.7 = \underline{2218.3\text{kN}}$

$R_{cC} = R_{bk}/\gamma_b + R_{sk}/\gamma_b = 1098.6/1.45 + 1119.7/1.3 = \underline{1619.0\text{kN}}$

Design Check

As $F_{cB} < R_{cB}$ (1987.0 < 2218.3) and $F_{cC} < R_{cC}$ (1509.6 < 1619.0), both Cases B and C are satisfied. As expected Case C is critical as F_{cC} is closer R_{cC} than F_{cB} is to R_{cB}.

7.5.5 Ultimate Bearing Resistance from Pile Driving Formulae

Not surprisingly, C7.6.3.4(1)P requires that, if pile driving formulae are to be used to assess the ultimate bearing resistance of piles for the design of piles to EC7, their validity shall be demonstrated by previous experimental evidence of good performance or by static load tests on the same type of pile, of similar length and cross section and in similar ground. Pile driving formulae shall only be used if the soil stratification has been determined [C7.6.3.4(2)P]. EC7 does not provide the same detailed information for determining the design bearing resistance from pile driving formulae as it does for deriving the bearing resistance from static load tests or from ground properties. Applying the method used to determine R_{cd} from static load tests, the pile driving formulae provide R_{cm} values which must be reduced by the ξ and γ factors in Tables 7.2 and 7.3 to obtain R_{cd}.

Driven piles can give false sets, particularly in silty soils due to increased resistance induced by dilation of the soil during shearing in the undrained condition. EC7 recommends that redriving be carried out in silty soils [C7.6.3.4(4)] and requires that the number of piles to be redriven is specified in the design [C7.6.3.4(3)P]; i.e. in the Geotechnical Design Report.

7.5.6 Ultimate Bearing Resistance from Wave Equation Analysis

The requirements in EC7 for assessing the ultimate bearing resistance of piles from wave equation analyses are similar to those for the assessment of R_{cd} from dynamic formulae. The validity of the wave equation analysis must be demonstrated by previous experience on static load tests on the same pile type, of similar length and cross section and in similar soil [C7.6.3.5(1)P]. It also required that the energy used is sufficiently high to allow interpretation of R_{cd} at the required strain level. If the dynamic load is too small, the full resistance of the soil will not be mobilised. It is noted in C7.6.3.5(3) that dynamic pile testing may supply a greater insight into the actual hammer performance and dynamic ground parameters, implying that such tests are more useful in providing information about the efficiency of the piling system for the particular site than on the pile bearing resistance.

As in the case of dynamic pile loads tests, it is important to determine the stratigraphy when using wave equation analysis to design piles. Hence C7.6.3.5(4)P has the requirement that wave equation analysis should normally only be used when the ground stratification has been determined by borings and field tests.

7.5.7 Settlement of Pile Foundations

The settlements of pile foundations shall be assessed for both ultimate and serviceability limit states [C7.6.4(1)P] and the values obtained compared against the limiting values for movements which are given §2.5.9.

Where ultimate limit states may occur in supported structures due to the settlement of piles before the ultimate bearing resistance is fully mobilised, the procedures outlined in C7.6.3 for determining the characteristic and design bearing resistances from load tests and ground properties shall be applied to the whole load-settlement curve [C7.6.4(2)P] to assess the pile settlement. This assessment shall involve the same treatment of downdrag as when assessing the bearing resistance and the same ξ and γ factors should be applied to the load-settlement curve.

For serviceability limit states, the settlements of pile foundations are calculated using characteristic loads and characteristic stiffness or consolidation parameters.

EC7 does not provide an indirect method for the design of pile foundations against the occurrence of a SLS based on the use of a large enough mobilisation factor in a ULS calculation to limit the deformations. Table 7.4 shows that the FOS values obtained when designing to EC7 against a ULS are similar to the FOS values traditionally used for pile design and which were chosen to avoid unacceptable settlements. Thus, in most cases, the settlements of piles designed for the ULS should be acceptable. Nevertheless, the pile settlements should be assessed.

7.6 Design of Piles in Tension

7.6.1 Tensile Resistance

The design of piles in tension must be consistent with the design of compression piles, and because of this, **C7.7** dealing with the design of such piles, only addresses the additional aspects which are to be considered **[C7.7.1(1)P]**. The requirement to be satisfied with regard to the design of piles in tension, given in **C7.6.3.1(1)P**, is that, for all ULS load cases and load combinations, there must be adequate safety against tensile resistance failure; i.e.

$$F_{td} \leq R_{td} \tag{7.3}$$

where:

F_{td} is the ULS design axial design tensile load;

R_{td} is ULS design tensile resistance of the pile foundation.

The design of piles in tension may be based on either load tests or ground test results, however design based on ground tests is only permitted when this method has been proven by load tests on similar piles, of similar length and cross-section, under comparable conditions **[C7.7.2.3(2)P]**. Thus EC7 essentially implies that a load test must be carried out when designing tension piles. This may be very restrictive and impractical in some situations. For example, it is generally impractical to test a tension pile when used in combination with a concrete plug, placed underwater, to prevent base uplift due to dewatering a deep excavation.

Two failure mechanisms must be considered, the pull out of individual piles and the uplift of the block of ground containing the piles **[C7.2.1(2)P]**. When considering the pull out of the ground containing the piles, the block may be modelled as a cone **[C7.7.2.1(3)]**. The group effect, which may reduce the vertical effective stress in the ground, and hence the shaft resistance, must be considered in the design **[C7.7.2.1(6)P]**. The severe adverse effect of cyclic loading on the tensile resistance of piles must also be considered **[C7.7.2.1(8)P]**.

7.6.2 Ultimate Tensile Resistance from Load Tests

The design of tension piles from load tests is similar to the design of compression piles from load tests in that the characteristic tensile resistance, R_{tk} is obtained from the measured resistance, R_{tm} by applying the correlation (reduction) factor, ξ that allows for the variability of the ground and the effects of pile installation; i.e.

$$R_{tk} = R_{tm}/\xi \tag{7.4}$$

The value of ξ depends on the number of pile test results as shown in Table 7.5. The values in Table 7.5 may revised when EC7 is issued as an EN. R_{tk} is the minimum value obtained applying the ξ values to the mean and lowest R_{tm} values.

The design tensile resistance, R_{td} is obtained by dividing R_{tk} by a partial factor, $_{st}$ (written as γ_m in EC7):

$$R_{td} = R_{tk}/\gamma_{st} \tag{7.5}$$

Number of load tests	1	2	>2
(a) Factor ξ on mean R_{tm}	1.5	1.35	1.3
(b) Factor ξ on lowest R_{tm}	1.5	1.25	1.1

Table 7.5: *Values of the correlation factor ξ to obtain the characteristic tensile resistance*

The γ_{st} value of 1.6, given in EC7, is for Case C. The Case B γ_{st} value is 1.0, but no value is given for Case A. By examining the ratio between the Case A and Case C partial factors on $\tan\phi'$, Simpson and Driscoll (1998) have deduced that the appropriate Case A γ_{st} value is 1.4, and this value has also been used by the authors.

Normally more than one pile should be tested, and about 2% should be tested where there is a large number of tension piles [**C7.7.2.2(3)**]. Note that load tests on trial tension piles should generally be taken to failure [**C7.5.2.1(2)**].

Example 7.4: Design of a Tension Pile using Load Tests

It is required to verify the design of a square precast concrete pile of side length 350mm which is driven 16m into the gravel to resist the uplift force due to flood conditions on the 10m diameter tank shown in Figure 7.3. The tank rests on dense sandy gravel with $\phi'_k = 40°$ and $\gamma = 22kN/m^3$. The area is liable to flooding to a depth of 5m above ground level. The weight of the tank is 3200kN. The ultimate tensile resistance of a trial pile located close by was 2100kN.

Design Situations
The design is checked for pull out of the pile from the ground mass. The possibility of failure involving uplift of the block of ground containing the pile is not considered to be relevant. Since this uplift example involves both buoyancy and the strength of the ground, the inequality $V \leq R$ is checked for Cases A, B and C.

Design Actions
When considering buoyancy, the actions in this example consist of the weight, W of the tank (favourable) and the uplift force, U on the tank and pile due to the water pressure (unfavourable). Although the buoyancy is due to flooding, the water pressure is considered to be a permanent action (§2.5.2). The weight of the pile is not included in the design actions when using the load test result because, as in Example 7.2, the tensile resistance obtained by applying a load to the top of a similar pile in the load test takes account of the pile weight. The characteristic loads are:

W_k = 3200kN

U_k = Area of tank*pressure on tank base + area of pile base*pressure on pile base
= $(\pi*10^2/4 - 0.35^2)*9.81*5.0 + 0.35^2*9.81*(5.0 + 16.0) = 3871.6kN$

Using the partial factor of 0.9 on the unfavourable action U_k for Case A, rather than the factor of 0.95 given in EC7, as in Example 6.5, the design loads are:

Case A: $F_{tA} = 1.0*U_k - 0.9*W_k = 3871.6 - 0.9*3200 = \underline{991.6kN}$
Case B: $F_{tB} = 1.35 (U_k - W_k) = 1.35 (3871.6 - 3200) = \underline{906.7kN}$
Case C: $F_{tC} = 1.0*U_k - 1.0*W_k = 3871.6 - 3200 = \underline{671.6kN}$

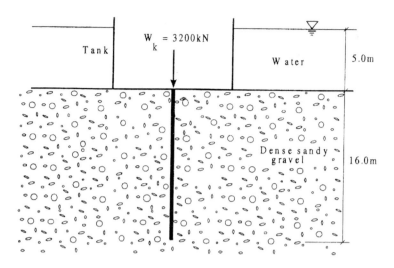

Figure 7.3: *Tension pile to resist uplift force on tank*

Design Resistances
The measured pile tensile resistance, R_{tm} = 2100kN, hence R_{tk} = 2000/1.5 = 1400.0kN. A partial resistance factor of γ_t = 1.6 is given for Case C and γ_t = 1.0 is implied for Case B, but no factor is provided in EC7 for Case A. , so a value of γ_t = 1.4, which is between the Case C value and unity (like the partial material factors for Case A), has been chosen:

Case A: $R_{tA} = R_{tk}/\gamma_t = 1400.0/1.4 = \underline{1000.0kN}$

Case B: $R_{tB} = R_{tk} = 1400.0 = \underline{1400.0kN}$

Case C: $R_{tC} = R_{tk}/\gamma_t = 1400.0/1.6 = \underline{875.0kN}$

Conclusion
As $F_{tA} < R_{tA}$ (991.6 < 1000), $F_{tB} < R_{tB}$ (906.7 < 1400.0) and $F_{tC} < R_{tC}$ (671.6 < 875.0), Cases A, B and C are satisfied and therefore the design of the pile is verified to resist flood conditions. Case A is the controlling case for the design situation in this example.

7.6.3 Ultimate Tensile Resistance from Ground Test Results
Calculation methods based on ground test results should only be used to determine the ultimate tensile resistance of a pile if they have been proved by load tests on similar piles, of similar length and in comparable soil conditions [**C7.7.2.3(1)**]. Situations do, however, arise where it is necessary to carry out preliminary pile design to determine the length of tension piles. Furthermore safety against failure due to uplift of the block of ground containing the pile needs to be checked and this will entail a calculation method involving the use of ground test results.

Example 7.5: Tensile Resistance using Ground Test Results

The tensile resistance of the square precast concrete tension pile in Example 7.4 is to be determined using ground test results. The soil is dense sandy gravel with $\phi'_k = 40°$ and $\gamma = 22kN/m^3$. The pile has a side length of 350mm and is driven 16m into the ground. The water table is at the ground surface.

Design Resistance
It should be noted that a calculation method based on ground test results must be verified by a tensile load test. For this example, the tensile resistance, which is the shaft resistance, is assumed to be given by $q_t = K_s\sigma'_{v0}\tan\delta$ per unit area of the shaft. For the driven pile in the dense gravel, $K_s = 1.5$ and it is assumed that $\delta = 0.9*\phi'_k = 36°$. Hence the calculated average tensile resistance per unit area is:

$$q_{tcal} = K_s\sigma'_{v0}\tan\delta = 1.5*\{(22.0 - 9.81)*(16.0/2)\}*\tan36 = 106.3kPa$$

Using the method given in §7.5.4 to determine the characteristic bearing resistance from ground test results, the characteristic tensile resistance per unit area is:

$$q_{tk} = q_{tcal}/1.5 = 106.3/1.5 = 70.9kPa$$

Hence the characteristic tensile resistance of the pile is:

$$R_{tk} = A_s*q_{tcal}/1.5 = (4*0.35*16.0)*70.9 = \underline{1588.2kN}$$

Using the partial factors in Example 7.4, the design tensile resistances are:

Case A: $R_{tA} = R_{tk}/\gamma_t = 1588.2/1.4 = \underline{1134.4kN}$

Case B: $R_{tB} = R_{tk} = \underline{1588.2kN}$

Case C: $R_{tC} = R_{tk}/\gamma_t = 1588.2/1.6 = \underline{992.6kN}$

Conclusion
As $F_{tA} < R_{tA}$ (991.6 < 1134.4), $F_{tB} < R_{tB}$ (906.7 < 1588.2) and $F_{tC} < R_{tC}$ (671.6 < 992.6), Cases A, B and C are all satisfied and the design is verified using the ground test results.

7.6.4 Vertical Displacement

The principles for assessing the vertical displacement of piles in tension with regard to the serviceability criteria are similar to those for piles in compression. Generally, according to C7.7.3(2), designs based on the ultimate tensile resistance will ensure that the vertical displacements satisfy the normal serviceability criteria, unless very severe criteria are given when a separate analysis of the displacements may be required.

7.7 Transversely Loaded Piles

When designing transversely loaded piles to EC7 the following inequality shall be satisfied:

$$F_{trd} \leq R_{trd} \tag{7.6}$$

where:

F_{trd} is the ULS axial design transverse load; and

R_{trd} is ULS design resistance against transverse loads, taking into account the effect of any compressive or tensile axial load.

The failure mechanisms to be considered in the case of a pile subjected to transverse loading are, for short piles, rotation or translation movement of the pile as a rigid body, and for long slender piles, bending failure of the pile accompanied by local yielding and displacement of the soil near the top of the pile. The ultimate transverse resistance can be determined from pile load tests [**C7.8.2.2**] or from ground test results and pile strength parameters [**C7.8.2.3**]. When using pile load tests, it is normally not necessary to continue the tests to failure. When using ground test results, the analysis should include the possibility of a broken pile in the ground. In the case of long slender piles, the transverse resistance of the ground may be modelled using an appropriate horizontal modulus of subgrade reaction. The degree of freedom at the connection with the structure shall be taken into account. The estimation of transverse displacements shall take into account the non-linearity of the soil stiffness, the flexural resistance of the individual piles, group effect, the fixity of the piles at the connection with the structure and the effect of load reversals.

7.8 Pile Load Tests

7.8.1 General

EC 7 considers two types of pile load test: trial pile load tests and working pile load tests. Trial piles are installed for test purposes only and before the design is finalised whereas working piles, as the name implies, form part of the foundation [**C7.4.1(2)**]. The load test procedure for both types of test is required to be such that conclusions can be drawn about the deformation, creep and rebound behaviour of the piles [**C7.5.2.1 (1)P**]. However, from trial pile tests, it shall also be possible to draw conclusions about the ultimate failure load [**C7.5.2.1(1)P**]. Pile load tests for designing tensile piles should generally be taken to failure [**C7.5.2.1(2)**].

Recommendations on where pile load tests should be carried out, where they should be located and the time required between construction and testing are provided in **C7.5.1**. The manner in which the load tests are to be carried out is to be specified in the Geotechnical Design Report [**C7.6.3.2(1)P**].

In the case of very large diameter piles it is often impractical to carry out a load test on a full size trial pile. The use of a smaller diameter pile is permitted by EC7 [**C7.6.3.2(3)**], provided the diameter of the trial pile is not less than 0.5 that of the working pile, provided the fabrication and installation of both are the same and provided the pile is instrumented such that the base and shaft resistances can be derived separately. The use of a smaller diameter pile for open ended piles should be used with caution owing to the influence of the diameter on the formation of a soil plug in the pile.

7.8.2 Static Load Tests

EC7 recommends in **C7.5.2.1(2)** that the testing procedure for trial piles be in accordance with the recommendations given in 'Axial pile loading test - Part 1: Static loading' published in the American Society for Testing Materials - Geotechnical Testing Journal (ASTM, 1985). This document was prepared by a sub-committee of the ISSMFE and represents an international consensus on good practice for static pile testing. However this document is not entirely consistent with EC7 and a European standard on static pile load testing is currently being prepared.

The various factors which need to be taken into account when deciding on the number of trial pile tests required to verify a pile design are listed in **C7.5.2.2(1)P**. These factors include the ground conditions and their variability, the geotechnical category of the structure, previous documented evidence, and the total number and type of pile. The ground conditions at the test site shall be investigated thoroughly and this investigation shall include all strata likely to contribute significantly to the pile deformation behaviour, at least to a depth of 5 times the pile diameter beneath the pile base, unless sound rock or very hard soil is encountered at a lesser depth [**C7.5.2.2(2)P**]. According to **C7.5.2.2(3)P**, the method of installation shall be fully documented.

The authors consider that in specifying the procedure for a compression pile load test, the loading system should preferably enable the inequality $F_{cd} \leq R_{cd}$ to be verified, without extrapolation. This means that the pile should be loaded to a test load of at least $R_{cC}* \xi * \gamma$ where R_{cC} is the required Case C compressive resistance and ξ and γ have the values given in Tables 7.2 and 7.3. For tension pile load tests, the test load should be at least $R_{tC}* \xi * 1.6$, where R_{tC} is the required Case C tensile resistance and the ξ values are given in Table 7.5. The test load values will vary with the number of piles and type of pile, however a test load greater than $2.25R_{cC}$ would be satisfactory for all compressive tests and a test load greater than $2.4 R_{tC}$ for all tensile tests. These test load values may change if the ξ and γ factors are modified when EC7 is revised as an EN.

EC7 does not provide guidance on the number of working piles to be tested, other than to mention that the number shall depend on the basis of the recorded findings during construction [**C7.5.2.3(1)P**]. The test load applied in working pile tests is to be at least equal to the load governing the design of the pile (adjusted for downdrag if this is expected [**C7.5.2.3(3)**]).

7.9   Structural Design of Piles

The requirements for the structural design of piles is covered by the clauses in **C7.9**. Piles should be design to accommodate all situations, including transport and, where applicable, driving [**C7.9(2)P**]. Slender piles passing through water or thick deposits of very weak soil need to be checked against buckling [**C7.9(4)P**]. This check is not normally necessary when piles are completely embedded in the ground unless the characteristic undrained shear strength is less than 15kPa [**C7.9(5)**]. The structural design of steel piles is covered in Section 5 of prENV 1993-5 (1997).

7.10   Supervision during Construction

C7.10(1)P requires that a detailed pile installation plan be prepared as the basis for the construction work. The information, such as pile type, location, cross-section, toe level or penetration resistance and known obstructions, which should be included in this installation plan is listed in **C7.10(2)**. The installation of piles shall be monitored and a record, signed by the supervisor and by the pile manufacturer, shall be kept [**7.10(3)P**] as the piles are installed. The items which should be recorded during pile installation are listed in **C7.10(4)**. These records shall be kept with the as-built drawings, for a minimum of 5 years [**C7.10(5)P**].

Additional investigations shall be undertaken if the site observations or inspection of records reveal uncertainties with respect to the quality of installed piles [**C7.10(5)P**]. These investigations shall include either redriving or pile integrity tests, in combination with geotechnical field tests on the ground adjoining the suspected piles and static load tests. Dynamic low strain integrity tests can be used for global evaluation of piles that may have suffered severe defects or that may have caused serious loss of strength of the surrounding soil. Such methods may not identify defects such as inadequate concrete quality or insufficient concrete cover and sonic tests, vibration tests or coring may also be required [**C7.10(8)**].

References

ASTM (1985) Axial pile loading test – Part 1: Static loading, *American Society for Testing Materials - Geotechnical Testing Journal*, 8(2), pp79-90.

Berezantsev V.G., Kristoforov V.S. and Golubkov V.N. (1961) Load-bearing capacity and deformation of piled foundations, *Proceedings, V International Conference on Soil Mechanics and Foundation Engineering*, Paris, 2, 11-15.

Frank R. (1997) Some comparisons of safety for axially loaded piles *Design of axially loaded piles-European Practice*, De Cock & Legrange (eds), pp39-46.

prENV 1993-5 (1997) *Eurocode 3: Design of steel structures – Part 5: Piling*, CEN Brussels.

Schuppener B. (1998) Internal communication to Working Group 1 of CEN TC 250/SC7, Eurocode 7: Geotechnical Design.

Simpson B. & Driscoll R. (1998) *Eurocode 7 – a commentary*, Construction Research Communications Ltd., London.

Chapter 8

Retaining Structures

8.1 Introduction

Section 8 of EC7 covers all types of walls and support systems in which structural elements are used to retain ground, similar material or water at a slope steeper than they would eventually adopt if no structure were present. It also covers anchorages used to support retaining structures, a subject which will probably be given a separate section when EC7 is revised as an EN.

Three main types of retaining structures are distinguished: gravity walls, embedded walls and composite retaining structures [**C8.1(2)**]. Gravity walls are those where the weight of the wall itself, sometimes including stabilised soil or rock masses, play a significant role in the support of the retained material. Embedded walls are relatively thin walls where the bending resistance plays a significant role in the support of the retained material while the weight of the wall is insignificant. Composite retained structures include walls composed of elements of the above two types. It is worth noting that earth structures reinforced by tendons, geotextiles or grouting, and structures with multiple rows of ground anchors or soils nails are included as composite retaining structures. At the time of publication there are three CEN standards which are relevant to retaining structures. These are Eurocode EC3 - Part 5: Design of steel structures – Piling (ENV 1993-5:1997) and the three standards on the Execution of special geotechnical works: EN 1537 – Ground anchors, EN 1538 - Diaphragm Walls and EN 12063 - Sheet pile walls.

This chapter briefly discusses the approach adopted in EC7 for the design of retaining structures and anchors, together with the requirements for this approach. The main differences between this approach and the more traditional methods for designing retaining structures are also discussed.

Three examples of wall designs are provided, a reinforced concrete cantilevered retaining wall, a cantilever sheet pile wall and an anchored sheet pile quay wall. An example of the design of a ground anchor is also included.

8.2 Design Approach

The selection of actions to EC7 must be in accordance with **C2.4.2**, that is the partial factors presented in Table 2.6 in §2.5.6 must be applied to the actions and ground properties for Cases A, B and C. As discussed in §2.5.6, earth pressures on both side of a retaining structure are treated as being derived from a single source [**C2.4.2(17)**], consequently all characteristics permanent earth pressures must be multiplied by 1.35 if the net effect is unfavourable. An alternative approach for Case B calculations is to use the partial load factor of 1.35 as a model factor [**C2.4.2 (17)**] and to apply this factor to the action effects. In this approach the earth pressures are calculated from the characteristic ground properties and the calculated bending moments and forces are multiplied by 1.35. When using this approach it is necessary to adjust the variable load by the ratio of 1.5/1.35 i.e. 1.11, to give the required safety margin. In many design situations both approaches give exactly the same answer.

8.3 Comparison of EC7 and Traditional Approaches

The implications of Cases B and C on the design a simple cantilevered retaining wall in a dry uniform coarse grained soil are illustrated in Figures 8.1a, 8.1b and 8.2a. The Case B and C design earth pressures are compared with those computed using the characteristic soil parameters. These pressures show that, in Case B, the design active and passive pressures are both increased compared with those computed using characteristic parameters, in Case C the active pressure is increased while the passive pressure is reduced, while in the traditional method, using an overall factor of safety on the passive pressure, the passive pressure is only decreased. These design earth pressures imply that Case C will always give a greater depth of embedment and will generally give a greater maximum bending moment as the passive resistance in front of the pile is less than that in Case B whereas the active pressure is similar in both cases.

As no partial factors are applied to water pressures in Case C, the relationship between the maximum bending moments in Case B and Case C is less easily compared and the former could be expected to be greater in certain circumstances.

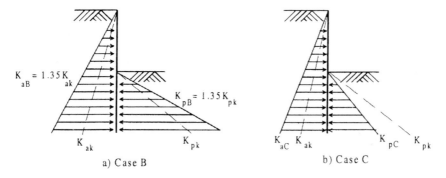

Figure 8.1: *Case B, Case C and characteristic earth pressures*

The partial factor approach in EC7 results in lower margins of safety in some design situations were large water pressures involved, when compared to the margins of safety implied in current design practice using lumped factors of safety. This arises because, as stated previously, no partial factor is applied to water pressures in EC7 as illustrated in Figure 8.2b. Farrell & Orr (1998) showed that in the extreme case of a high lateral water force on a sheet pile wall, the traditional factor of safety of a structure designed to Case B and Case C can be below 1.3. This low safety margin is acceptable if the most unfavourable ground water conditions are used, as required by EC7 [**C2.4.2(10)P**]. Apart from the extreme case of high water pressures, the safety margin provided by the partial factors in designs to EC7 is similar to the safety margins provided using traditional methods.

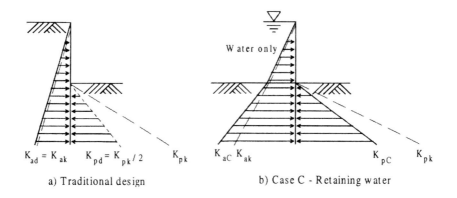

a) Traditional design   b) Case C - Retaining water

Figure 8.2: *Traditional design earth pressures and Case C earth pressures when only water retained*

Schuppener et al. (1998) proposed that an alternative approach be taken whereby partial factors are applied to the resisting forces, not to the soil strength parameters. This approach will probably be included in the EN version of EC7 as an alternative method of analysis that may be called Case C2, as discussed in §2.5.6.

8.4 Limit States

C8.2(1)P requires that a list of the limit states be compiled. As a minimum the limit states listed in Table 8.1 are required to be checked. Combinations of these limit states must also be considered [**C8.2(3)P**]. When considering the limit states for gravity structures, the principles of **Section 6** on Spread Foundations, should be applied, with particular care being taken to account for bearing resistance failure below the base of a wall under loads with large eccentricities and inclinations [**C8.2(4)**].

Limit states to be considered	Retaining structure type	Checked
Loss of overall stability	All types	
Failure of structural element e.g. wall, anchor, strut, connection	All types	
Combined failure in ground and in structural element	All types	
Movements of the retaining structure which may cause collapse or affect the appearance or efficient use of the structure, nearby structures or services	All types	
Unacceptable leakage through or beneath the wall	All types	
Unacceptable change to the flow of groundwater	All types	
Bearing resistance failure of the soil below the base	Gravity and composite	
Failure by sliding at the base of the wall	Gravity and composite	
Failure by toppling of the wall	Gravity and composite	
Failure by rotation or translation of the wall or parts thereof	Embedded	
Failure by lack of vertical equilibrium	Embedded	

Table 8.1: *Limit states to be considered in the design of retaining structures*

8.5 Actions

The actions listed in **C2.4.2** must be considered in the design of retaining structures to EC7 [**C8.3.1(1)P**]. If the unit weight of the backfill to a retaining structure is to be checked during construction, the type of test and the number of tests to be carried out are to be detailed in the GDR. In practice this means that a conservative estimate should be given to the weight of the backfill to avoid having to check the unit weight of the fill during construction. When considering surcharges, allowance should be made for the increased earth pressures which arise from repeated surcharge loadings such as those induced by crane rails [**C8.3.1.2(2)**]. The lateral pressure from these sources may be significantly greater than that applied by static loading.

The effect of contaminants, salinity and mud on the unit weight of water must be considered as these can significantly affect the unit weight [**C8.3.1.3(1)P**]. Design for wave and wave impact forces shall be selected on the basis of local site conditions [**C8.3.1.4(1)P**]. Design forces shall take into account the energy absorbed by the retaining system on impact [**C8.3.1.6(1)P**]. The impact of an ice floe colliding with a retaining structure shall be calculated on the basis of the compressive strength of the ice and on the thickness of the ice [**C8.3.1.6(3)P**]. The design values for ice forces from a sheet of ice covering water shall be calculated taking into account the initial temperature of the ice before warming begins, the rate at which the temperature increases and the thickness of the ice [**C8.3.1.7(3)P**]. Special precautions shall be taken to prevent ice lenses forming in the ground behind retaining structures [**C8.3.1.7(4)P**].

The supporting forces caused by prestressing operations are to be considered as actions [**C8.3.1.5(1)P**]. The selected design values must take into account the effect of overstressing the anchor, the effect of relaxation and the effect of abnormal temperature differences over time and space [**C8.3.1.7(1)P**]. These temperature effects are particularly important when designing struts and props [**C8.3.1.7(2)**].

8.6 Geometrical Data

C8.3.2(1)P requires that the design values for geometrical data be determined in accordance with the principles in **C2.4.5** (see §2.5.4). When determining the design level for ground surfaces, special consideration must be given to the possibility of unplanned excavations or possible scour in front of the retaining structure [**C8.3.2.1(1)P**]. In certain circumstances, such as an embedded wall in a dense gravel with a high ϕ'_k, a relatively small excavation in front of the wall can have a major effect on the overall stability (Orr, 1999). **C8.3.2.1(2)** therefore recommends that in ULS calculations, where the stability of a retaining wall **depends on the passive resistance** of the ground in front of the retaining wall, the ground level of the passive soil should be lowered by an amount Δ_a. For embedded cantilever walls, $\Delta_a = 10\%$ of its height and for a supported wall $\Delta_a = 10\%$ of the height beneath the lowest support with Δ_a limited to a maximum of 0.5m.

8.7 Design and Construction Considerations

According to **C8.4(1)P**, earth retaining structures must be designed against ultimate and serviceability limit states using a combination of the approaches mentioned in **C2.1(7)** (see §2.2.4) which are design by calculations, by prescriptive measures, by experimental models and load tests and by the observational method. The use of the observational method is specifically mentioned in **C8.4(2)** as being appropriate for the design of retaining structures as it is sometimes difficult to design such a wall before construction commences because of complex ground/structure interaction effects that often occur.

Earth retaining structures must deflect in order to mobilise the soil resistance and the resulting ground movements could cause damage to nearby structures and services. It is difficult to predict these movements, however **C8.4(2)** notes that in dense granular soils or in firm fine soils, these movements should not give rise to an SLS in the surrounding structures if the retaining structure is designed and constructed in accordance with EC7. Thus an SLS calculation is not normally required in those circumstances. This is not the case where retaining structures are formed in highly overconsolidated soils where the release of high initial horizontal stresses can cause deflections over a large area.

The design of earth retaining structures must consider the sequence, feasibility and practical aspect of construction [**C8.4(3)P**]. The design should guard against the brittle failure of retaining structures; i.e. sudden collapse without conspicuous preliminary deformations [**C8.4.(4)**]. In practice this should be complied with by a critical assessment of the deformation characteristics of the structure. This is particularly relevant in the case of reinforced structures relying on the tensile strength of fibres or ground anchors.

Where the safety or serviceability depends on the successful performance of a drainage system, as is the case in many retaining structures, the consequences of failure of the system with regard to life and the cost of repair must be considered [C8.4(5)P]. In such cases EC7 requires confirmation that the drainage system works, either by:

a) specifying a maintenance programme in the GDR and making provision in the design for access to the drainage system in order to carry out this maintenance;

b) demonstrating, both by comparable experience and by assessment of the water which emerges, that the drainage system will operate effectively without maintenance; for example, the water emerging from drains may be checked for fines.

8.8   Determination of Earth and Water Pressures

As noted in §2.5.2, the term 'earth pressure' in EC7 includes the pressure from soils and weathered rocks **and** the groundwater [C8.5.1(2)]. Earth pressure does not include the pressure from other granular material, such as those stored in silos, and reference is made in C8.5.1(2) to ENV 1991-4: Eurocode 1 – Part 4: Actions on silos, for the determination of pressures from such materials.

The magnitude of earth pressures on a structure depends on the way the structure moves relative to the soil and on the level of strain which is acceptable and which may take place at the limit state under consideration. C8.5.1(1)P requires that these movements and limits be taken into account and lists some of the factors which must be considered in C8.5.1(3)P. Guidance is given in C8.5.1(4) on the amount of mobilised wall friction, δ, and wall adhesion, α. For a completely smooth wall $\delta = 0$ and $\alpha = 0$, while for a completely rough wall $\delta = \phi'$ and $\alpha = c$, where c is the 'apparent cohesion'. A concrete or steel sheet piled wall supporting sand or gravel normally may be assumed to have $\delta = k \phi'$ and $\alpha = 0$. The value of ϕ' should not exceed the critical state angle of friction of the ground due to the disturbance at the ground wall interface and k should not exceed the value of 2/3 for precast concrete or steel sheet piling but which it may assume a value of unity for concrete cast against soil. A steel sheet pile in clay under undrained conditions normally should be assumed to have $\delta = 0$ and $\alpha = 0$ immediately after driving. Regeneration may take place over a period of time.

The magnitudes and directions of the design earth pressures are required to be calculated using the partial factors given in Table 2.6. It should be noted that the design value of an earth pressure at a ULS is generally different from its value at an SLS. These two values are determined from two fundamentally different calculations. Consequently, when expressed as an action, earth pressures cannot be characterised by a single characteristic value [C8.5.1(6)]. EC7 requires that for retaining structure calculations for rock masses, the discontinuities and their orientation be considered [C8.5.1(7)P]. Where relevant, swelling pressures from cohesive soils[C8.5.(8)P] shall also be taken into account.

The following five different types of earth pressure are considered in the subsections of **C8.5**:

a) at rest earth pressure [**C8.5.2**];
b) limit values of earth pressure [**C8.5.3**];
c) intermediate values of earth pressure [**C8.5.4**];
d) earth pressure due to compaction[**C8.5.5**];
e) water pressure [**C8.5.6**].

a) At rest earth pressure
This is the earth pressure when no movement of the wall relative of the ground takes place. The actual value of the earth pressure in this design situation depends on the stress history of the soil [**C8.5.2(1)P**]. At rest pressure will normally apply when the movement of the structure is less than $5 \times 10^{-4} * H$ for normally consolidated soil, where H is the height of the retaining wall [**C8.5.2(2)**]. For horizontal ground $K_0 = (1 - \sin\phi') * (R_{oc})^{0.5}$ where R_{oc} is the overconsolidation ratio. This formula is not recommended for high values of R_{oc}. For ground shelving upwards at an angle $\beta \leq \phi'$, the horizontal component of the effective earth pressure σ_{h0} may be related to the effective overburden pressure by the ratio $K_{h0} = K_0 (1 + \sin\beta)$. The direction of the earth pressure force may then be assumed to be parallel to the ground surface.

b) Limit values of earth pressure
The limit earth pressure values are those which apply when the movement of the wall is sufficient to mobilise either the active or passive pressure [**C8.5.3(1)P**]. EC7 gives the following guidelines for the movement required to achieve the active state for non-cohesive ground of at least medium dense consistency:

*   rotation about the top < 0.002H;
*   rotation about the toe < 0.005H;
*   translative motion < 0.001 H.

EC7 includes an informative **Annex G** which shows how limiting earth pressure values may be calculated. This Annex includes graphs of the active and passive earth pressure coefficients based on Caquot, Kerisel & Absi (1973). The graphs are not convenient for spreadsheets or for numerical formulation of design problems as it is more convenient if the active and passive earth pressure coefficients are expressed as a function. The authors have found that the Müller-Breslau equation for the active pressure coefficients for frictional cohesionless soil, as given in Clayton et al (1996), gives active earth pressure coefficient values that are close to those in EC7, and are on the safe side. However to date no similar equation for the passive pressure coefficients is available. **Annex G** includes a numerical procedure for determining earth pressures. However, this procedure includes some typographical errors and the authors have found it difficult to use.

c) Intermediate values of earth pressure
Intermediate earth pressure values occur when there is insufficient movement to mobilise the limiting values [**C8.5.4(1)P**]. Intermediate earth pressure values may be calculated using empirical rules, spring constant methods, finite element, etc. Generally, in the design of retaining structures to EC7, it is assumed that walls can

move sufficiently so as to develop the limiting active and passive earth pressures. Intermediate earth pressure values are relevant for situations where wall movement is prevented as, for example, in the cases of stiff basement wall.

d) Earth pressures due to compaction
The compaction of fill behind a wall induces earth pressures greater than the active earth pressure. The magnitude of the pressure depends on the applied energy, thickness of the compacted layer and the fill height. These additional earth pressures must be considered in design [**C8.5.5(1)P**]. Compaction procedures must be applied during construction to ensure that these pressures do not damage the structure [**C8.5.5(3)P**]. Appropriate methods for estimating these compaction pressures are given by Broms (1971) and Ingold (1979).

e) Water pressure
As there is no partial factor on water pressures for Case C, the selection of the design water pressure is an important aspect in the design of retaining structures. Water pressures shall be selected in accordance with **C2.4.2**, together the requirements of **C8.4(5)P**, which relate to the verification of drainage provisions if these are essential for the stability of the wall. As discussed in §2.5.2, **C2.4.2(10)P** requires that the most unfavourable values of water pressures and seepage forces which could occur in extreme circumstances be taken for limit states with severe consequences, i.e. ultimate limit states. **C2.4.2(19)P** permits the use of methods of determining the design value of ground and groundwater other than by use of partial factors, however the level of safety should be consistent with that implied by the partial factors in Table 2.6. Case C is frequently the critical case when high water pressures are involved and, as the partial factor on water pressure is unity for Case C, this clause implies that an increased safety margin on the water pressure is not required for this case.

Non-steady, as well as steady state, seepage pressures must be considered where sudden changes in water level may occur [**C8.5.6(4)P**]. EC7 requires that for structures retaining silts and clays, the water pressure behind the wall shall be assumed to correspond to a water table at a level no lower than the top face of the material with low permeability, unless a reliable drainage system is installed or infiltration is prevented [**C8.5.6(3)P**]. Where no special drainage or flow prevention measures are taken, the possibly effect of water filled tension or shrinkage cracks shall be considered [**C8.5.6(5)P**]. For these circumstances in retained cohesive soil, the design total earth pressure should normally not be less than the pressure of water increasing hydrostatically from zero at the ground surface [**C8.5.6(6)**].

8.9   Ultimate Limit State Design

The design of earth retaining structures shall be checked at the ultimate limit state using the design actions and design situations appropriate to that limit state [**C8.6.1(1)P**] as discussed in §8.5. The possible limit states listed in Table 8.1 and the failure modes illustrated in Figures 8.1 to 8.6 of EC7 should be considered. Overall stability shall be verified using the principles of **Section 9**, [**C8.6.2(1)P**].

C8.6.1(4)P requires that ULS calculations establish that equilibrium can be achieved using design actions and design strengths. This clause also requires that compatibility of deformations shall be considered when assessing design strengths. **C8.6.5(1)P** emphasises the former point and requires that vertical equilibrium be verified for embedded walls. This requirement frequently goes unchecked in current practice. Vertical equilibrium shall be achieved by altering the wall friction values on one side of the wall [**C8.5.3(3)P**]. It is generally found that approximate vertical equilibrium can be achieved for sheet piled walls using $\delta/\phi' = 0.5$ on the active side and $\delta/\phi' = 0.67$ on the passive side. These are wall friction ratios which were recommended by Terzaghi as mentioned in Clayton et al. (1996). **C8.6.4(3)P** requires that, for embedded walls, the magnitude and direction of shear stresses between the soil and the wall be consistent with the relative vertical displacement which would occur in the design situation.

8.6.1(5)P requires that the design values of the ground parameters represent the most adverse situation. Generally these design values are based on lower values of the strength or stiffness. However there are exceptional cases, such as the case of backfill to an integral bridge, where the upper values of these parameters would represent the most adverse condition. In such design situations, the characteristic value must be a cautious estimate of the upper characteristic value of the appropriate strength parameter. Ultimate limit states can occur in either the short (undrained) or long term (drained) condition for fine-grained soils and **C8.6.1(7)P** requires that both be considered for such soils. The safety of walls subjected to differential water pressure against hydraulic instability (erosion, piping or heave) must be checked [**C8.6.1(8)P**]. Hydraulic instability is discussed in §2.5.6.

EC7 has sections which relate to the ultimate limit state design for foundation failure of gravity walls, for rotational failure of embedded walls and for vertical failure of embedded walls. The principles behind these sections have been discussed above and will not be repeated.

The structural design of retaining structures is discussed in **C8.6.6**. The deformations required to mobilise the strength in the ground must be consistent with those required to mobilise the structural strength [**C8.6.6(3)P**]. The design should consider the reduction in strength in structural elements due to cracking, plastic hinges etc. in accordance with the other structural Eurocodes [**C8.6.6(4)**]. The design should also allow for the loss of strength in the ground, such as by dilation in granular soils or by the formation of slickensides in clay. **C8.6.6(4)** also introduces the concept of a model factor γ_{sd} without giving a value for this factor or any guidance about it use. The use of a model factor in Case B has been discussed in §8.2 and its use is illustrated in Example 8.3.

The requirements for the design of anchors against a ULS of pull-out, including the failure modes that need to be considered, are presented in **C8.6.7**. These requirements are mostly of a general nature, apart from the recommendation in **C8.6.7(2)** that no allowance for wall friction be made when calculating the passive resistance of 'dead men' anchors in pull-out calculations.

In practice it is difficult to model the Case B design situation correctly using numerical methods, such as finite elements, by applying partial load factors as the actions are not known at the start and it would be necessary to apply the partial factors to the unit weight in accordance with Table 2.6. These difficulties can be overcome by using the model factor approach as described in §8.2. In this approach the analysis is carried out using the characteristic ground properties and then the computed forces and moments, the action effects, are multiplied by the appropriate partial factor. It is necessary, as noted in §8.2, to adjust the variable load to allow for the model factor.

8.10 Serviceability Limit State Design

C8.7.1(1)P requires that the serviceability limit state be checked for each design situation. Limiting values of the allowable displacement of the structure and of the surrounding ground must be established [C8.7.2(1)P]. The deformations in and around retaining structures can be a complicated ground/structure interaction problem and therefore can be difficult to analyse. This is recognised in the EC7 as C8.7.2(2)P requires that a cautious estimate of the distortion and displacement be made on the basis of comparable experience. This estimate must consider the effect of construction. C8.7.2(3)P requires that a more detailed investigation and analysis be carried out if the initial cautious estimates of displacements exceed the limiting values. C8.7.2(4)P lists the following design situations where a detailed investigation and analysis must be carried out if the initial cautiously estimated displacements are greater than 50% of the limiting values:

- where nearby structures and services are sensitive to displacement;
- where the wall retains more than 6m of soil of low plasticity or 3m of soil of high plasticity;
- where the wall is supported by soft clay within its height or beneath its base;
- where comparable experience is not well established.

When using models, such as finite elements or spring stiffnesses, to calculate displacements, the material properties should be appropriate for the strain level and the calculation should be calibrated against comparable experience [C8.7.2(6)].

Design earth pressures for checking the SLS of structural elements shall be derived using characteristic values of all soil parameters [C8.7.4(1)P]. When checking the SLS of structural elements, the derived design earth pressures should also take into account the deformation of the structure at its serviceability limit state [C8.7.4(2)]. These earth pressure will not necessarily be the active and passive pressures.

It should be noted that the allowance for overdig in C8.3.2.1(1)P is not required when carrying out an SLS check. The estimation of deformation of retaining structures and the ground around such structures is normally carried out using numerical models of various forms, generally on commercially available software. Some guidance on the movements to be expected from certain types of retaining walls can be obtained from the case histories and empirical methods outlined in Clough and O'Rourke (1990).

Example 8.1: Reinforced Concrete Cantilever Retaining Wall

A concrete cantilever retaining wall with compacted backfill, which is to form part of an artificial lake, has the dimensions shown in Figure 8.3. The backfill is a compacted, coarse, free draining material with $\gamma = 20kN/m^3$, $c'_k = 0$ and $\phi'_k = 40°$. The wall is founded on a low plasticity fine grained glacial till with $\gamma = 22kN/m^3$, $c_{uk} = 150kPa$, $c'_k = 0$ and $\phi'_k = 37°$. The concrete unit weight is $23.4kN/m^3$. In this example the ultimate limit states of a) bearing resistance failure, b) failure by sliding along the wall base, and c) failure by toppling are considered. The serviceability limit state is also addressed.

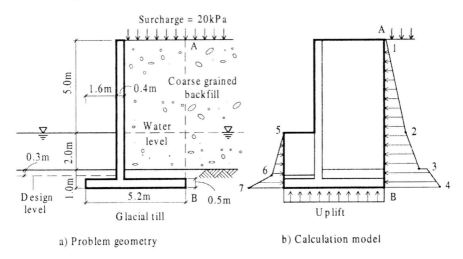

a) Problem geometry   b) Calculation model

Figure 8.3: *Reinforced concrete cantilever retaining wall*

Bearing resistance

Calculation model and geometric profile

This example shows how the actions required to design the wall against bearing resistance failure and sliding, using the methods discussed in Chapter 6, are determined. The actions are assessed by assuming that the retaining wall and the soil above the stem behave as a unit, i.e. that the wall has a virtual back (AB in Figure 8.3) vertically above the heel. It can be shown that the length of the heel is such that the compaction stresses would not affect the virtual back (Clayton et al, 1996), hence full Rankine active pressure conditions, i.e. horizontal earth pressures parallel to the ground surface, are assumed to act on AB.

After consideration of likely erosion and overdig in front of the wall, an allowance of 0.3m has been adopted to account for possible reduction in the ground level in front of the retaining wall. It should be noted that the recommendation in **C8.3.2.1(2)** that the ground level should be lowered by 10% of the height of the wall does not apply in this case as the stability of the wall depends principally on the weight of the structure and the base resistance and not on the passive resistance in front of the wall.

Soil	Parameter	Drained		Undrained	
		Case B	Case C	Case B	Case C
Backfill ($\gamma = 20\text{kN/m}^3$)	$\phi'(^\circ)$ K_a K_P	40.0 0.22 -	33.9 0.284 -	Not applicable	Not Applicable
Glacial till ($\gamma = 22\text{kN/m}^3$)	c_u (kPa) $\phi'(^\circ)$ K_a K_P	- 37.0 0.25 4.0	- 31.1 0.32 3.0	150 - 1.0 1.0	107.14 - 1.0 1.0

Table 8.2: *Ground parameter values and earth pressure coefficients*

Design earth pressures for ultimate limit state calculations
The most important aspect in the analysis of earth retaining structures is the determination of earth pressures. The earth pressure distribution for this example is illustrated in Figure 8.3, with the locations where there is a change in the pressure gradient, or where the pressure is used in the calculation, denoted by identification numbers (ID). The design ϕ' values and the design earth pressure coefficients, K for drained and undrained conditions for Case B and C are given in Table 8.2.

The values in Table 8.2 are used to calculate the design earth pressures at the ID points. These calculations are presented in Table 8.3 for drained conditions for Cases B and C and in Table 8.4 for undrained conditions for Cases B and C. Where the ID points coincide with a change in strata, the pressure relating to the lower stratum is denoted by the ID number with a superscript [+]. Note that in the Case B calculations, for the calculation method adopted, the partial factors are included in the pressures given Table 8.3 and 8.4.

As discussed in §2.5.6, for Case B, both the active and passive earth pressures, including the water pressures, should be treated as being derived from a single source. C2.4.2(17) requires that both be multiplied by the appropriate Case B partial factor. In this example the partial factor has been applied to the pressures, rather than to the forces. Note that, as also discussed in §2.5.6, the partial factors are applied to the net vertical force in the Case B calculations.

Design forces and moments for checking the ultimate bearing resistance
The following is a sample calculation, for this example, of the design vertical force, V_C and the disturbing moment M_C about the centre of the base from the earth pressures given in Table 8.3 for the drained (long-term) conditions for Case C, which is the critical situation.

$$
\begin{aligned}
V_C &= 1.0*\{23.4*0.4*7.5 + 23.4*0.5*5.2 + 20*7*3.2 + 22*0.5*3.2 + \\
&\quad 9.81*1.6*2.3 + 22*0.2*1.6 - 3*9.81*5.2\} \\
&= 70.2 + 60.84 + 448 + 35.2 + 36.1 + 7.04 - 153.0 \\
&= \underline{504.3\text{kN/m}}
\end{aligned}
$$

ID	Case B		Case C	
	Earth pressure calculations	kPa	Earth pressure calculations	kPa
1	1.5*0.22*20	6.6	1.3*0.284*20	7.4
2	1.5*0.22*20 + 1.35*0.22*20*5	36.3	1.3*0.284*20 + 1.00*0.284*20*5	35.8
3	1.5*0.22*20 + 1.35*0.22*(20*7 - 9.81*2) + 1.35*9.81*2	68.8	1.3*0.284*20 + 1.0*0.284*(20*7 - 9.81*2) + 1.0*9.81*2	61.2
3$^+$	1.5*0.25*20 + 1.35*0.25*(20*7 - 9.81*2) + 1.35*9.81*2	74.6	1.3*0.32*20 + 1.0*0.32*(20*7 - 9.81*2) + 1.0*9.81*2	66.5
4	1.5*0.25*20 + 1.35*0.25*(20*7 + 22*1 - 9.81*3) + 1.35*9.81*3	92.0	1.3*0.32*20 + 1.0*0.32*(20*7 + 22*1 - 9.81*3) + 1.0*9.81*3	80.2
5	0	0	0	0
6	1.35*9.81*2.3	30.5	1.0*9.81*2.3	22.6
6$^+$		30.5		22.6
7	1.35*4*(9.81*2.3+22*0.7 - 3*9.81) + 1.35*9.81*3	85.8	1.0*3*(9.81*2.3+22*0.7 - 3*9.81) + 1.0*9.81*3	55.0

Table 8.3: *Earth pressure calculations and earth pressures for drained conditions*

ID	Case B		Case C	
	Earth pressure calculations	kPa	Earth pressure calculations	kPa
1	1.5*0.22*20	6.6	1.3*0.284*20	7.4
2	1.5*0.22*20 + 1.35*0.22*20*5	36.3	1.3*0.284*20 + 1.00*0.284*20*5	35.8
3	1.5*0.22*20 + 1.35*0.22* (20*7 - 9.81*2) + 1.35*9.81*2	68.8	1.3*0.284*20 + 1.0*0.284*(20*7 - 9.81*2) + 1.0*9.81*2	61.2
3$^+$	1.5*1.0*20 + 1.35*(1.0*20*7 - 2*150)	(-186.0) 0	1.3*1.0*20 + 1.0*1.0*20*7 - 2*107.14	(-48.3) 0
4	1.5*1.0*20 + 1.35*1.0*(20*7 + 22*1 - 2*150)	(-156.3) 0	1.3*1.0*20 + 1.0*1.0*(20*7 + 22*1) - 2*107.14	(-26.3) 0
5	0	0	0	0
6	1.35*9.81*2.3	30.5	1.0*9.81*2.3	22.6
6$^+$	1.35*9.81*2.3 + 1.35*(2*150)	435.5	1.0*9.81*2.3 + 1.0*2*107.14	236.8
7	1.35*1*9.81*2.3 + 1.35*(1.0*22*0.7 + 2*150)	456.3	1.0*1.0*9.81*2.3 + 1.0*22*0.7 + 2*107.14	252.3

Table 8.4: *Earth pressure calculations and earth pressures for undrained conditions*

Design situation	Case	P (kN/m)	M (kNm/m)	e = M/V
Drained	B	504.3	242.7	0.48
Undrained	B	657.4	106.1	0.16
Drained	C	504.3	248.1	0.49
Undrained	C	657.4	161.5	0.25

Table 8.5: *Forces, moments and eccentricities of forces*

$$
\begin{aligned}
M_C \ = \ & 1.0*\{70.2*(-0.8) + 60.84*0 + 448*(-1) + 35.2*(-1) + 36.1*1.8 + \\
& 7.04*1.8 - 153.0*0\} + 7.4*5*5.5 + 0.5*(35.8 - 7.4)*5*4.67 + 35.8*2*2 + \\
& 0.5*(61.2 - 35.8)*2*1.67 + 66.5*1*0.5 + 0.5*(80.2 - 66.5)*1*0.33 - \\
& 0.5*22.6*2.3*1.47 - 22.6*0.7*0.35 - 0.5*(55.0 - 22.6)*0.7*0.23 \\
\ = \ & \underline{248.1\text{kNm/m}}
\end{aligned}
$$

The vertical force and moment per metre length of wall and the eccentricity for all the design conditions are given in Table 8.5. These forces and eccentricities can be used in calculations to assess bearing resistance as outlined in §6.5.2.

Sliding resistance

Design situation

The calculation model, geometric profile and design earth pressure parameters are as discussed above. The treatment of the actions is similar to that adopted above, in that, for Case B, the net vertical force due to the downward weight of the soil above the heel and the uplift force due to the water pressure beneath the base are treated as favourable actions. For the limit state of sliding, as discussed in §6.5.5, **C6.5.3(2)P** requires that the design horizontal thrust (H_d) does not exceed the design horizontal resistance ($R_d = S_{d+}E_{pd}$); i.e.

$$H_d \leq S_d + E_{pd}$$

where S_d is the sliding resistance of the base of the wall and E_{pd} the design earth pressure resistance on the side of the base which can be mobilised with the displacement appropriate to the limit state being considered and is available throughout the life of the structure. In this example it is assumed that the full E_{pd} is available.

The Case B calculation for H_d and $S_{d+}E_{pd}$ for drained conditions is:

$$
\begin{aligned}
H_B \ = \ & 6.6*5 + 0.5*(36.3-6.6)*5 + 36.3*2 + 0.5*(68.8 - 36.3)*2 + 74.6*1 + \\
& 0.5*(92.0 - 74.6)*1 \\
\ = \ & \underline{295.7\text{kN/m}}
\end{aligned}
$$

$$
\begin{aligned}
S_{d+}E_{pd} \ = \ & 504.34*\tan37° + 0.5*30.5*2.3 + 30.5*0.7 + 0.5*(85.8-30.5)*0.7 \\
\ = \ & 455.8\text{kN/m}
\end{aligned}
$$

Conclusion

As $H_d < S_d + E_{pd}$ (295.7 < 455.8), there is adequate safety against failure due to sliding for Case B and drained conditions. The results of the calculations for all the

Design situation	Case	H_d	$S_d + E_{pd}$	Assessment
Drained	B	295.7	455.8	$H_B < S_B + E_{pB}$
Undrained	B	212.4	1127.2	$H_B < S_B + E_{pB}$
Drained	C	278.4	357.4	$H_C < S_C + E_{pC}$
Undrained	C	205.0	754.3	$H_C < S_C + E_{pC}$

Table 8.6: *Assessment of the different Cases for stability against sliding*

design situations, taking the undrained resistance along the base as equal to the undrained shear strength of the till where appropriate, and their assessment are presented in Table 8.6.

Failure by toppling
Generally a toppling limit state will only be critical when a retaining structure is on rock or very stiff soil as bearing resistance failure is generally the critical mechanism. However the analysis of toppling traditionally has been carried out as a matter of routine when designing retaining walls. This mechanism can be analysed using the same approach as that adopted above to determine the forces and moments for a bearing resistance calculation except the moments are taken about the toe rather than about the centreline of the wall. For example, taking Case B for the drained situation and noting that the partial factors for the unfavourable actions have been included in the earth pressures:

$$
\begin{aligned}
M_{TB} &= 6.6*5*5.5 + 0.5*(36.3 - 6.6)*5*4.67 + 36.3*2*2 + 0.5*(68.8 - \\
&\quad 36.3)*2*1.67 + 74.6*1*0.5 + 0.5*(92 - 74.6)*1*0.33 - \\
&\quad 0.5*30.5*2.3*1.47 - 30.5*0.7*0.35 - 0.5*(85.8 - 30.5)*0.7*0.23 \\
&= \underline{704.4 \text{kNm/m}}
\end{aligned}
$$

$$
\begin{aligned}
M_{RB} &= 1.0*\{23.4*0.4*7.5*1.8 + 23.4*0.5*5.2*2.6 + 20*7*3.2*3.6 + \\
&\quad 22*0.5*3.2*3.6 + 9.81*1.6*2.3*0.8 + 22*0.2*1.6*0.8 - 3*9.81*5.2*2.6\} \\
&= \underline{1660.7 \text{ kNm/m}}
\end{aligned}
$$

where M_{TB} is the overturning moment and M_{RB} is the resisting moment for Case B. The values obtained for M_{TB} and M_{RB} show that for Case B and drained conditions, the toppling requirement $M_{TB} < M_{RB}$ is satisfied as 704.4 < 1660.7. Similar calculations may be carried out for the other design situations.

Serviceability limit state
No specific requirements regarding the wall displacements are given in this example. The aesthetic appearance of the wall should be maintained. It is considered that the serviceability limit states will not be critical if the ultimate limit states are satisfied as the EC7 partial factors give a similar safety margin to those used in traditional designs, which have been found to be satisfactory with regard to serviceability limit state requirements, unless high water pressures are involved.

It is important to note that the design earth pressures used for ultimate limit state calculations are not necessarily those appropriate for serviceability limit state calculations as the required deformations may not be mobilised.

Example 8.2: Cantilever Sheet Pile Cofferdam in Layered Estuarine Soil

A cantilever sheet pile wall, which is part of a cofferdam in a river, is driven into a sandy gravel in order to retain water and soft silt as shown in Figure 8.4. The characteristic properties of the soft silt are $\gamma_k = 18kN/m^3$, $c'_k = 0$, $\phi'_k = 33°$ and $c_{uk} = 10kPa$, and those of the sandy gravel are $\gamma_k = 20kN/m^3$, $c'_k = 0$ and $\phi'_k = 37°$. It is assumed that the cofferdam will be required for a reasonably long period so that both undrained and drained drainage conditions need to be considered. In this example the ULS of rotation of the sheet pile is analysed to determine the length of the pile and the maximum bending moment for use in assessing the structural strength of the sheet pile sections. The serviceability limit state is also addressed.

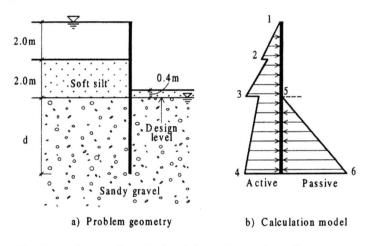

a) Problem geometry   b) Calculation model

Figure 8.4: *Cantilevered sheet pile cofferdam in layered estuarine soil*

Characteristic ground profile
The stability of the wall relies on passive resistance on the soil in front of the wall, hence the design ground level in front of the wall is reduced by $\Delta_a = 10\%$ of the wall height of 3.6m, i.e. by 0.4m to allow for unplanned excavations, dredging, etc.

Design earth pressures parameters for ultimate limit state calculations
It is necessary to select the appropriate groundwater pressures for the two design cases: for the gravel only in the undrained case and for the soft silt and gravel in the drained case. The groundwater pressures depend on the relative permeability of the soft silt in comparison with that of the gravel, on the consolidation characteristics of the soft silt, and on various geohydraulic conditions. For the purposes of this example, it is assumed that ¾ of the hydraulic head is lost uniformly through the silt and that the remaining head is lost uniformly along the penetration depth of the sheet pile.

The coefficients of active and passive earth pressure are obtained from the Müller-Breslau formula and from the graphs in Annex G of EC7 respectively. The

Soil	Parameter	Drained		Undrained	
		Case B	Case C	Case B	Case C
Soft silt	c_u (kPa)	0.0	0.0	10.0	7.14
	$\phi'(^\circ)$	33	27.45	0	0
	K_a	0.27	0.33	1.0	1.0
	K_P	-	-	1.0	1.0
Sandy gravel	c_u (kPa)	-	-	Not applicable	Not applicable
	$\phi'(^\circ)$	37.0	31.1		
	K_a	0.23	0.29		
	K_P	9.2	5.7		

Table 8.7: *Ground parameter values and earth pressure coefficients*

value of δ/ϕ' is taken as 0.5 on the active side of the wall and as 2/3 on the passive side, which is in line with Terzaghi's recommendation as discussed in §8.9. The design ground parameter values and earth pressure coefficients are given in Table 8.7. Note that partial factors given in EC7 are for both persistent and transient situations. According to **C2.4.2(14)P**, less severe (i.e. lower) partial factor values may be used for transient (i.e. temporary) situations if these can be justified on the basis of the possible consequences but no guidance or less severe values are provided. It is possible that some guidance on the partial factor values for transient situations may be provided in the EN version.

Calculation model
The stability of the wall is analysed using Blum's simplified method in which the sheet pile is considered to rotate about the toe. It is assumed that active pressure conditions are developed on the rear of the wall and passive pressures in front of the wall. The calculated penetration of the sheet pile is increased by 20%, in accordance with conventional practice, to allow for the development of passive resistance behind the heel if this is not checked using horizontal equilibrium. The length of the wall is determined so that the disturbing moment (M_D) does not exceed the resisting moment (M_R); i.e. $M_D \leq M_R$. Vertical equilibrium is assumed to be satisfied (§8.9).

The derivation of the ultimate limit state design earth pressures for the drained conditions are presented in Table 8.8. As the earth pressures are unfavourable actions, in the Case B calculations the appropriate partial factors have been applied to the pressures, rather than to the forces, which would be more correct.

The locations of changes in pressure gradient are denoted by ID values in Table 8.8, as explained in Example 8.1. The depth of penetration of the sheet pile, which is to be determined, is designated by the symbol d in Table 8.8.

Calculations
The calculation model for this example involves taking moments about the toe. As there are no variable loads in this example and as both the active and passive pressures are multiplied by 1.35 while the soil strengths are unfactored in Case B, there is no margin of safety in this calculation model for Case B. The length of the pile, therefore, will always be determined by Case C unless there is an unusually

ID	Case B		Case C	
	Earth pressure calculations	kPa	Earth pressure calculations	kPa
1	0	0	0	0
2	1.35*(2*9.81)	26.5	2*9.81	19.6
2⁺		26.5		19.6
3	1.35*{0.27*(2*9.81 + 2*18 - 9.81) + 9.81}	29.9	0.33*(2*9.81 + 2*18 - 9.81) + 9.81	24.9
3⁺	1.35*{0.23*(2*9.81 + 2*18 - 9.81) + 9.81}	27.5	0.29*(2*9.81 + 2*18 - 9.81) + 9.81	23.1
4	1.35*{0.23*(2*9.81 + 2*18 + d*20 - 9.81*(d + 0.5)) + 9.81*(d + 0.5)}		0.29*(2*9.81 + 2*18 + d*20 - 9.81*(d + 0.5)) + 9.81*(d + 0.5)	
5	0	0	0	0
6	1.35*{9.2*(d*20 - 9.81*(d + 0.5)) + 9.81*(d + 0.5)}		5.7*(d*20 - 9.81*(d + 0.5)) + 9.81*(d + 0.5)	

Table 8.8: *Earth pressure calculations and earth pressures for drained conditions*

ID	Case B		Case C	
	Earth pressure calculation	kPa	Earth pressure calculation	kPa
1	0	0	0	0
2	1.35*2*9.81	26.49	2*9.81	19.6
2⁺	1.35*{1*2*9.81 - 2*10}	(-0.5) 0	1*2*9.81-2*7.14	5.3
3	1.35*{1*(2*9.81 + 2*18) - 2*10}	48.1	1*(2*9.81+2*18)-2*7.14	41.3
3⁺	1.35*{0.23*(2*9.81 + 2*18 - 9.81) + 9.81}	27.5	0.29*(2*9.81 + 2*18 - 9.81) + 9.81	23.1
4	1.35*{0.23*(2*9.81 + 2*18 + d*20 - 9.81*(d + 0.5)) +9.81*(d + 0.5)}		0.29*(2*9.81 + 2*18 + d*20 - 9.81*(d + 0.5)) + 9.81*(d + 0.5)	
5	0	0	0	0
6	1.35*{9.2*(d*20 - 9.81*(d + 0.5)) + 9.81*(d + 0.5)}		5.7*(d*20 - 9.81*(d + 0.5)) + 9.81*(d + 0.5)	

Table 8.9: *Earth pressure calculations and earth pressures for undrained conditions*

high proportion of variable load. This is confirmed in the resulting lengths and bending moments computed for the different cases shown in Table 8.10. The maximum bending moment, however, is obtained from Case B for drained calculations. It should be noted that exactly the same results are obtained for Case B calculations using the model factor approach as discussed in §8.2.

Case	Length of sheet pile wall (m)	Maximum bending moment (kNm/m)
B drained	7.0	221*
C drained	8.0	203
B undrained	6.9	184
C undrained	8.0	200

* determined using Case B length. The sheet pile in this case can be re-analysed using the design length which would allow redistribution of the stresses and a lower bending moment, if required.

Table 8.10: *Calculated sheet pile wall lengths and bending moments*

Design length

Sample calculation
The following is a sample stability calculation for the critical design situation of Case C for drained conditions with a sheet pile length of 8.0m so that d = 8.0 − 3.6 − 0.4 = 4.0m (the active and passive earth pressures at the toe of the sheet pile are 70.73kPa and 248.81kPa respectively):

M_{DC} = 0.5*19.6*2*6.68 + 19.6*2*5.01 + 0.5*(24.9 - 19.6)*2*4.68 + 23.1*4.01*2.005 + 0.5*(70.73 - 23.1)*4.01*1.34

= 665.8 kNm/m

M_{RC} = 0.5*248.81*4.01*1.34

= 668.5kNm/m

Conclusion
As M_D < M_R (665.8 < 668.5), Case C is satisfied and hence the design length is satisfactory. As noted in the section entitled calculation model, the depth of penetration is increased by 20% to allow for the development of passive resistance below the point of rotation if this is not separately calculated, thus the design length would be 3.6 + 0.4 + 4.0*1.2 = 8.8m.

Design bending moment
The critical design bending moment is obtained in the Case B calculation (with the length determined for Case B). As the length is controlled by Case C, an analysis could be carried out using this length and the Case B actions, as the Case B resistances would not be fully mobilised. This would allow redistribution of the earth pressures and a reduction in the Case B bending moment from the calculated value of 221kNm/m. Such an analysis would require a ground-structure interaction model using finite elements or spring stiffnesses, or similar methods.

Serviceability Limit State
As in the Example 8.1, no specific requirements regarding the wall displacements are given in this example. The aesthetic appearance of the wall should be maintained. As noted in Example 8.1, the serviceability limit state should not be critical if the ultimate limit states are satisfied.

Example 8.3: Sheet Pile Quay Wall

This example involves the design of the anchored sheet pile quay wall shown in Figure 8.5a. The wall is to retain 10m of coarse gravel, with properties $\gamma_k = 22kN/m^3$ and $\phi'_k = 35°$, and penetrates a silty sand, with properties $\gamma_k = 20kN/m^3$ and $\phi'_k = 32°$. The design tidal lag, with weep holes constructed in the sheetpiles, is assumed to be 0.3m. The quay wall is to be designed to support a uniformly distributed surcharge of 20kPa behind the wall. The quay is in a marine environment and the unit weight of water is $10kN/m^3$. The model factor method is adopted for Case B calculations.

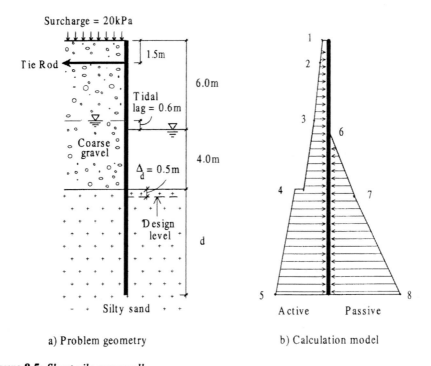

a) Problem geometry   b) Calculation model

Figure 8.5: *Sheet pile quay wall*

Design situation

All the soils are coarse grained, only drained conditions need be considered. It is assumed that the permeability of the backfill is significantly greater than that of the underlying silty sand, consequently all the hydraulic head is lost through the latter soil. It will also be assumed that this head is lost evenly along the length of the sheet pile in the silty sand.

Soil	Parameter	Drained		Undrained	
		Case B	Case C	Case B	Case C
Granular backfill	$\phi'(^\circ)$	35	29.26	Not applicable	Not applicable
	K_a	0.25	0.31		
	K_P	-	-		
Silty sand	$\phi'(^\circ)$	32.0	26.58	Not applicable	Not applicable
	K_a	0.28	0.35		
	K_P	6.1	4.2		

Table 8.11: *Ground parameter values and earth pressure coefficients*

Design geometry
The stability of the wall relies on the passive resistance, hence the design ground level in front of the wall is reduced by 10% of the height up to a maximum of 0.5m to allow for unplanned excavations, scour, etc. This gives a design height of 10.5m.

Design actions
The tidal lag is an action caused by groundwater and is therefore treated as a permanent action [**C2.4.2(16)P**]. A tidal lag of 0.3m is traditionally used for such quay walls with weep holes (EAU, 1980), however the authors have increased this to 0.6m because of the requirement in **C2.4.2(10)P** that ULS design water pressure shall be the most unfavourable value which could occur in extreme circumstances.

Design earth pressure coefficients
The active pressure coefficients have been determined using the Müller-Breslau formula and the passive pressure coefficients from the charts in EC7 as described in Example 8.2. The design coefficients are given on Table 8.11.

Ultimate limit state design earth pressures
The model factor method is used in this example for Case B calculations in which the characteristic earth pressures are calculated and the resulting moments and shear forces in the wall are then multiplied by 1.35. It is necessary to multiply the variable loads by $1.5/1.35 = 1.111$ when using this method to allow for the higher partial factor on the variable loads and hence provide the appropriate safety margin. The penetration depth of the sheet pile, which is the unknown in the calculations, is designated d. The earth pressures at the changes in gradient are given in Table 8.12, using the ID system explained in Example 8.1, and plotted in Figure 8.5b.

Calculation model
The design approach adopted is the 'Free Earth' method where the sheet pile is considered to be a rigid beam rotating about the anchor. It is assumed that active pressure develops on the back of the wall and passive pressure on the front. This calculation model determines a) the length of the sheet pile wall, b) the design bending moment and c) the design anchor force.

ID	Case B		Case C	
	Earth pressure calculations	kPa	Earth pressure calculations	kPa
1	1.111*0.25*20	5.6	1.3*0.31*20	8.1
2	1.111*0.25*20 + 1.0*{0.25*(22*1.5)}	13.8	1.3*0.31*20 + 1.0*{0.31*(22*1.5)}	18.3
3	1.111*0.25*20 + 1.0*{0.25*(22*5.4)}	35.3	1.3*0.31*20 + 1.0*{0.31*(22*5.4)}	44.9
4	1.111*0.25*20 + 1.0*{0.25*(22*10 - 4.6*10) + 4.6*10}	95.1	1.3*0.31*20 + 1.0*{0.31*(22*10 - 4.6*10) + 4.6*10)}	108.0
4$^+$	1.111*0.28*20 + 1.0*{0.28*(22*10 - 4.6*10) + 4.6*10}	100.9	1.3*0.35*20 + 1.0*{0.35*(22*10 - 4.6*10) + .6*10)}	116.0
5	1.111*0.28*20 + 1.0*{0.28*[22*10 + 18*d - (d + 4.6)*10 + 0.6*d*10/(2d - 0.5)] + (d + 4.6)*10 - 0.6*d*10/(2d - 0.5)}		1.3*0.35*20 + 1.0*{0.35* [22*10 + 18*d - (d + 4.6)*10 + 0.6*d*10/(2d - 0.5)] + (d + 4.6)*10 - 0.6*d*10/(2d - 0.5)}	
6	0	0	0	0
7	1.0*10*4.5	45	10*4.5	45
8	1.0*{6.1*[4.5*10 + (d-0.5)*18 - (d + 4.0)*10 - 0.6*(d - 0.5)*10/(2d - .5)] + (d + 4.0)*10 + 0.6*(d - 0.5)*10/(2d - 0.5)}		1.0*{4.2*[4.5*10 + (d - 0.5)*18 - (d + 4.0)*10 - 0.6*(d - 0.5)*10/(2d - 0.5)] + (d + 4.0)*10 + 0.6*(d - 0.5)*10/(2d - 0.5)}	

Table 8.12: *Earth pressure calculations and earth pressures*

The length of the wall is determined such that the design disturbing moment about the anchor (M_{DT}) does not exceed the design resisting moment (M_{RT}); i.e. $M_{DT} \leq M_{RT}$. The length of the sheet pile wall is determined by Case C rather than Case B. This is because, in a Case B calculation, the earth pressures on both sides of the wall, which are the dominant actions in this situation, are treated as permanent actions and multiplied by 1.35, and as there is no factor on the soil strength, there is no margin of safety in a Case B calculation. The length of the pile calculated is 17.1m and this is verified by the following calculation (the total active and passive pressures at the wall toe are 204.34kP and 321.73kPa respectively).

Vertical equilibrium
The downward component of the active force was 236.0kN/m while the upward component of the passive force was 219.7kN/m, thus there was a slight imbalance which can be ignored. This imbalance in vertical equilibrium is rectified by a slight alteration of the δ/ϕ' ratio on the active or passive side as required by **C8.5.3(3)P**. In this example, a reduction of δ/ϕ' on the active side from 0.5 to 0.46 satisfies vertical equilibrium without any significant alteration in the maximum moments and forces. The values computed using $\delta/\phi' = 0.5$ on the active side and 0.67 on the

Case	Calculated length (m)	Maximum bending moment (kNm/m)	Anchor force (kN/m)
Case B	14.8	1.35*774.3 = 1045.3	1.35* 220.0 = 297.0
Case C	17.1	1252	321

Table 8.13: *Calculated sheet pile wall lengths, bending moments and anchor forces*

passive side are retained in this example. As noted in §8.9, the use of δ/ϕ' values of 0.5 and 0.67 on the active and passive sides respectively usually results in near vertical equilibrium.

Design length
The stability of the wall for the Case C earth pressures in Table 8.12 is analysed as follows:

M_{DTC} = - 8.1*1.5*1.5/2 - 0.5*(18.3 - 8.1)*1.5*1.5/3 + 18.3*3.9*3.9/2 + 0.5*(44.9
    - 18.3)*3.9*3.9*2/3 + 44.9*4.6*(3.9 + 4.6/2) + 0.5*(108.0 -
    44.9)*4.6*(3.9 + 4.6*2/3) + 116.0*7.1*(8.5 + 7.1/2) + 0.5*(204.34 -
    116.0)*7.1*(8.5 + 7.1*2/3)

    = <u>16627.2kNm/m</u>

M_{RTC} = 0.5*45*4.5*(4.5 + 4.5*2/3) + 45*(7.1 - 0.5)*(9 + (7.1 - 0.5)/2) +
    0.5*(321.73 - 45)*(7.1 - 0.5)*(9 + (7.1 - 0.5)*2/3)

    = <u>16649.5kNm/m</u>

Conclusion
As $M_{DT} < M_{RT}$ (16627.2 < 16649.5), Case C is satisfied and hence the design length is satisfactory.

For Case B, application of the model factor of 1.35 to the maximum wall bending moment and anchor force calculated using characteristic earth pressures (with the adjustment for the variable load) gave exactly the same results as calculated using the method in Example 8.2 where the partial factors are applied to the earth pressures before calculating the bending moment and anchor force.

Case C gave the maximum length, as expected, and the maximum bending moment and anchor force for this example. The bending moment can be used directly in the calculation of the sheet pile section in accordance with EC3-Part 5 or reduction factors can be applied to allow for redistribution of moments as detailed in EAU(1980). Where conditions are favourable for such moment redistribution, the anchor force must be increased to allow for arching. A conservative approach in the determination of the design anchor force would be appropriate in order to avoid a brittle structure as recommended in **C8.4(4)**.

Serviceability limit state design
As with the previous examples, no SLS requirements have been specified. It is therefore considered that the wall deformations will be acceptable when the USL requirements are satisfied.

8.11 Anchorages

8.11.1 Introduction
EC7 covers any type of anchorage used to support a retaining structure by transmitting a tensile force to a load bearing formation of soil or rock [(**8.8.1(1)P**]. Anchorages include grouted anchors with a free length, reinforced concrete or steel anchors with a free anchor length, anchors without a free length (soil nails) and screw anchors. Both temporary and permanent anchors are considered. The execution (construction) of anchors is covered in EN 1537 Ground Anchors, however this standard does not cover screw anchors, mechanical anchors, soil nails, expander anchors or deadman anchors. This section discusses the design approach presented in EC7 but it is likely that revisions will be introduced in the EN.

8.11.2 Anchorage Design and Construction
The design of anchorages to EC7 is similar to that of piles in that partial factors are applied to characteristic resistances rather than to characteristic soil properties. As with piles, both Cases B and C should be considered, although the partial factors are only given for Case C because the design is controlled by Case C. A complication which arises in the design of anchorages is that the design anchor load may be controlled by SLS requirements, for example prestressing of the anchorage may be required to limit lateral displacements of an anchored wall. In such a calculation there is no partial factor on the load and therefore the SLS design load may exceed the ULS design load obtained using factored loads and ground properties. This may result in an unsatisfactory margin of safety against pull-out under the SLS load compared with normal practice. This aspect needs to be considered in design.

Anchorages must be designed to take into account all circumstances during the foreseeable design life of the anchorage, including corrosion and creep [**C8.8.2(1)P**]. Anchorages which are going to be in use for more than two years shall be designed as permanent anchorages [**C8.8.2(3)P**].

To check an anchorage against the occurrence of a ULS, **C8.8.2(4)P** requires that the following three failure mechanisms be analysed:
- failure of the tendon or anchor head or failure of bonding at internal surfaces;
- failure of the anchorage at the tendon/grout or grout/ground interface;
- overall stability failure of the structure, including the anchorages.

At present, EC7 provides no requirements regarding the serviceability limit state design of anchorages. It is likely that the EN version will include requirements and guidance relating to additional limit states, for example loss of anchor force by excessive displacement of the anchor head or by creep and relaxation and also failure or excessive deformation of parts of the structure due to the anchor force.

Steel tendons and bars used for anchorages shall be designed according to the principles of EC3-5 [**C8.8.2(6)P**]. The current requirement that the minimum free anchor length be about 5m [**C8.8.2(7)**] will probably be deleted in the EN version and replaced with a more general recommendation that the free anchor length must be appropriate to the design situation.

The connection between the tendon and the wall shall be able to adjust to the movements occurring in service and permanent anchors shall be protected against

Number of assessment tests	1	2	>2
(a) Factor ξ on mean R_{cm}	1.5	1.35	1.3
(b) Factor ξ on lowest R_{cm}	1.5	1.25	1.1

Table 8.14: *Conversion factors for anchorages*

corrosion for their full length, including the anchor head [**C8.8.3 (1)P & (2)P**]. A mechanical protection is also recommended for permanent anchors to cover handling installation etc. Indicative levels of compounds in the ground are given in **C8.8.3(3)** which, if exceeded, could cause ground water to be aggressive to concrete or hardened cement grout.

8.11.3 Ultimate Limit State Design

EC7 requires that the load carrying capacity of anchors be evaluated from the results of load tests and from local experience [**C8.8.4(1)P**], although this requirement will probably be limited to grouted anchors in the EN. For example, it is generally unnecessary to carry out a load test on a dead man anchor. The characteristic anchorage resistance R_{ak}, must be determined [**C(8.8.5(4)P**] from the measured ultimate anchor capacity R_{am} determined from assessment tests using:

$$R_{ak} = R_{am}/\xi \qquad (8.1)$$

where the conversion values, ξ, which vary with the number of tests, are given in Table 8.14. The ultimate characteristic anchorage resistance is the minimum value obtained using both (a) and (b) ξ values in table 8.14. The anchorage resistance, R_{am} is the lowest of loads corresponding to the relevant failure mechanism listed in §8.11.2 and the creep limit load.

The design resistance, R_{ad} is then obtained from the equation:

$$R_{ad} = R_{ak}/\gamma_m \qquad (8.2)$$

where γ_m is 1.25 for temporary anchors and 1.5 for permanent anchorages. The equivalent overall factor of safety on the resistance is therefore $\gamma_m *\zeta$ and is 1.875 and 2.25 for a single test on a temporary and permanent anchor respectively.

This design approach may be altered in the EN to $P_d \le R_{ad}$ where P_d is the design value of the anchor load and R_{ad} is the design value of the pull-out resistance. It is proposed to differentiate between the anchor force A, which is the anchor force required for the design of the supported structure, and the anchor load, P which is the tensile force applied to the anchor. The characteristic value of each is equal, i.e. $P_k = A_k$, however the design anchor load P_d is obtained from the characteristic anchor force A_k by applying a partial load factor, γ_p to the latter; i.e. $P_d = \gamma_p A_k$.

It is also possible that the conversion factor ξ will be omitted, resulting in R_{ak} being equal to the lowest value recorded in anchor assessment tests (in accordance with EN 1537) and that a partial factor γ_a will be introduced to take into account unfavourable deviations of the material properties, so that $R_{ad} = R_{ak}/\gamma_a$.

The current proposal is for γ_p to be 1.35 for Case B and unity for Case C and for serviceability limit checks and for γ_a to be 1.1 for Case B and 1.5 for Case C.

Using this approach the factor of safety for Cases B & C will be $\gamma_p*\gamma_a$ and are thus 1.485 and 1.5 respectively. Case B calculations are required for the design of the anchor. These partial factors may be altered before being published in the EN.

8.11.4 Serviceability Limit State Design

Although not expressly stated in EC7, the SLS of the structure being anchored and of structures which are influenced by the anchor force must also be considered. The effect of an anchor may be considered as a prestressed elastic spring.

8.11.5 Anchorage Testing

Anchorage testing is an integral part of anchor design to EC7. Two types of tests are considered, assessment tests and acceptance tests [**C8.8.4(1)P**]. Assessment tests are carried out in advance of the main contract or on selected working anchorages during construction to assess the suitability of the system to provide the necessary anchorage resistance [**C8.8.4(2)P**]. EN 1537 gives further information on testing of anchors and classifies two types of assessment tests, namely investigation tests and suitability tests, along with acceptance tests. Investigation tests establish in advance of the installation of the working anchors the ultimate resistance (R_{am}) of an anchorage at the grout/ground interface as well as the creep characteristics and apparent tendon free length. Suitability tests confirm, for a particular design situation, the ability of an anchor to sustain a proof load, P_p as well as determine the creep characteristics up to this proof load and the apparent tendon free length.

C8.8.4(3)P requires that acceptance tests demonstrate that each anchorage has the capacity to carry the design load and **C8.8.6(1)P** requires that an acceptance tests be carried out on all grouted anchorages. These tests confirm the ability of the anchors to take a proof load, determine the creep and load loss characteristics at the serviceability limit state, if required, and the apparent tendon free length. Details of the testing procedures for these tests are given in EN 1537.

Investigation tests must be carried out unless there is comparable documented experience of the successful performance in relation to performance and durability (EN 1537). At least one assessment test shall be carried out for each distinct ground conditions unless comparable experience exists [**C8.8.5(1)P**] . On large projects the number of assessment tests for each ground condition should be at least 1% for temporary anchorages where failure will have few serious consequences and 2% for permanent or temporary anchors where the consequences could be serious [**C8.8.5(2)P**]. EN 1537 requires at least three suitability tests. The anchorage should be loaded to failure (R_a) or to a proof load (P_P) which takes account of the strength limitations of the anchorage material properties. The proof load is defined in EN 1537 as the maximum test load to which the anchor is subjected. Acceptance tests should be carried out on all anchorages.

8.11.6 Supervision of Construction and Monitoring

An anchorage installation plan must be available on site containing the technical specification related to the anchorage system to be used [**C8.8.7(1)P**]. The installation plan should include, as appropriate, the information listed in **C8.8.7(2)**. The installation of all anchorages must be monitored and records made at the site and as the anchorages are installed and a signed record kept of each anchorage [**C8.8.7(3)P**].

Example 8.4: Design of a Permanent Anchor

A permanent anchorage is required to support a characteristic permanent load of 550kN. The anchorage resistances recorded in two assessment tests were 1254kN and 1451kN.

Design loads

P_B = 1.35*550 = 742.5kN
P_C = P_k = 550kN

The characteristic load is used in Case C design, but as the Case B load may depend on the reaction of the structure to a particular critical loading conditions, it is not necessarily obtained by applying partial factors to the load determined using characteristic loads on the structure. This is reflected in the use of anchor loads determined from structural analysis.

Design resistance

Two assessment tests were carried out giving R_{am} values of 1254kN and 1451kN. The lowest R_{am} value is therefore 1254kN and the average value is (1254 + 1451)/2 = 1352.5kN. With reference to Table 8.14, the characteristic ultimate anchorage resistance R_{ak} is the lowest of 1254/1.25 = 1003.2kN and 1352.5/1.35 = 1001.9kN; therefore R_{ak} = 1001.9kN. Applying the partial factors of 1.0 and 1.5 for permanent anchors for Cases B and C respectively, the design anchorage resistances are:

R_{aB} = 1001.9kN
R_{aC} = 1001.85/1.5 = 667.9kN

Conclusion

As $P_B < R_{aB}$ (742.5kN < 1001.9kN) and as $P_C < R_{aC}$ (550kN < 667.9kN), the design is satisfactory.

References

Broms B. (1971) Lateral pressure due to compaction of cohesionless soils, *Proc. IV Int. Conf. on Soil Mechanics and Foundation Engineering*, Budapest, 373-384.

Caquot A., Kerisel J. and Absi E. (1973) *Tables de buté et de poussé*. Gauthier-Villars.

Clayton C.R.., Militistky J. & Woods R.I., (1996*) Earth pressure and earth retaining structures*, Blackie Academic & Professional.

Clough G. W. and O'Rourke T.D (1990), Construction induced movements of in situ walls, *Proc. Conf. On Design & Performance of Earth Retaining Structures*, Cornell, ASCE, pp 439-471.

EAU (1980) *Recommendations of the committee for waterfront structures*, Wilhelm Ernst & Sohn, Berlin-Munich.

Farrell E.R. and Orr T.L.L. (1998) Safety of retaining walls with high water loadings when designing to Eurocode 7 using partial factors, *Ground Engineering*, October, pp37-40.

Ingold T.S. (1979) The effects of compaction on retaining structures, *Géotechnique*, 29(3), 265-283.

Orr T.L.L. (1999), Selection of characteristic values and partial factors in geotechnical designs to Eurocode 7, *Computers and Geotechnics*, Special issue on Reliability in Geotechnics.

Chapter 9

Embankments and Slopes

9.1 Introduction

Section 9 of EC7 provides the requirements for the design of both embankments and slopes. Although this section is primarily concerned with the requirements for the design of unsupported and unreinforced embankments and slopes in soil, it also provides the requirements for the design of rock slopes. While the design requirements are provided, there is little guidance on the analytical methods to be used, particularly in the case of rock slopes. According to **C9.1(1)P** the provisions of this section do not apply to dykes and dams. What was intended by this was that the provisions of this section, while applicable, may not be sufficient for the design of dykes and dams because of the higher risk associated with water retaining structures. Such structures would normally be classified as GC3 and therefore, according to **C2.1(5)**, additional investigations and analyses are required.

It is anticipated that, in the EN version of EC7, this section will be split into two sections, one entitled Site Stability, which will cover the overall stability of a site as well as the deformation behaviour of soil and rock masses, and another section entitled Embankments, which will provide the requirements for the design of embankments, dykes and dams.

9.2 Limit States and Design Situations

The different limit states that, according to **C9.2(1)P**, need to be considered in the design of slopes and embankments are given in Table 9.1 in the form of a checklist. It should be noted that deformations of slopes and embankments causing either structural damage or loss of serviceability in adjacent structures, roads and services

Number	Limit states	Checked
1	Loss of overall stability	
2	Bearing resistance failure in the case of embankments	
3	Failure due to internal erosion	
4	Failure due to surface erosion or scour	
5	Failure due to hydraulic uplift	
6	Deformations (excessive ground movements), including creep of slopes that cause structural damage to adjacent structures, roads or services	
7	Deformations (ground movements) which cause loss of serviceability to adjacent structures, roads or services	
8	Rockfalls	
9	Surface erosion	

Table 9.1: *Limit states to be considered in the design of slopes and embankments*

are both included in this list. It should also be noted that, of the nine limit states listed in Table 9.1, three are concerned with water:

- failure due to internal erosion;
- failure due to surface erosion or scour;
- failure due to hydraulic uplift pressures.

The fact that three of the limit states for slopes are concerned with water (both groundwater and surface water) demonstrates the important influence of water on the stability of slopes. Indeed many occurrences of slope instability arise following periods of heavy rainfall. It is critically important, therefore, that the effect of water is correctly taken into account and that the most unfavourable groundwater conditions within slopes and free water levels in front of slopes are selected when assessing slope stability. EC7 provides some guidance on the selection of water levels; for example, in the case of embankments along water, the most unfavourable, i.e. ultimate limit state, hydraulic conditions, are normally the highest possible groundwater level and rapid drawdown of the free water level in front of the slope [**C9.3(4)**].

The porewater pressure distribution in a slope is affected both by anisotropy and variability of the soil, and this should be taken into account. This is particularly important in the case of large embankments where the soil tends to be denser towards the base of the embankment due to the higher normal stresses at the base resulting in a lower permeability and hence reduced flow.

When specifying the design situations for a slope or embankment, a number of factors need to be considered [**C9.3(2)P**]. These factors include construction processes, such as excavations in front of a slope, which can have a great influence on the stability of a slope. The factors that need to be considered when selecting the design situations for a slope or embankment have been collected together in Table 9.2. One of the factors listed in Table 9.2 is continuing movement (often referred to as progressive movement) of existing slopes. This factor is included because of the importance, when assessing the suitability of a site or designing a new slope, of

Number	Factors to be considered	Checked
1	Excavations in front of a slope	
2	Vibrations caused by rock blasting	
3	Vibrations caused by piling	
4	Structures that are expected to be placed on a slope or embankment after its completion	
5	Effects of new slope on existing work	
6	Effects of previous or continuing movement of existing slopes	
7	Effects of overtopping, waves or rain on slopes and crests of embankments	
8	Effects of temperature causing cracking	
9	Effects of frost, freezing and thawing	
10	Effects of vegetation	
11	Effects of variations in pore water pressure and free water levels	
12	Animal activities clogging drains or digging holes	
13	Possibility of failure of drains, filters or seals	
14	Possible range of anisotropy or variability of the soil	
15	Dynamic effects caused by earthquake ground movements	

Table 9.2: *Factors to be considered when selecting design situations*

identifying any previous or continuing movement of existing slopes. Previous movement of slopes will usually be indicated by features such as inclined trees, heave zones at the toe of a slope, uneven slope surfaces and irregular vegetation on slope surfaces.

As well as affecting the pore water pressure distribution in a slope or embankment, variability of the soil can determine the mechanism of failure. For example if thin bands of weak soil exist in a slope, failure may occur by the soil sliding along these weak bands. For this reason comprehensive investigation of the ground and groundwater conditions is essential for all stability assessments.

9.3 Design Considerations

The various factors and options referred to in **C9.4** that should be considered in the design and construction of slopes and embankments are shown in Table 9.3. It should be noted that the first factor listed in Table 9.3 is the need to take account of experience of existing slopes in similar ground. This emphasises the importance of visiting the site of a slope and observing the stability and condition of the neighbouring slopes.

In the case of unstable existing slopes, the possibility of stabilising these by modifying the geometry, for example by reducing the slope, should be considered. In the construction of embankments it is most important to ensure that soil is placed in zones or layers and compacted so as to achieve the design performance requirements as discussed in §5.2.5. Embankments constructed on soft ground are a

Number	Factors and Options	Checked
1	Experience of slopes and cuttings in similar ground	
2	Installation of drains to control groundwater	
3	Modification of the geometry of existing slopes	
4	Construction of embankments in layers	
5	Measurement of settlements and lateral displacements	
6	Measurement of pore pressures	
7	Protection of surfaces exposed to erosion	

Table 9.3: *Factors and options to be considered in the design of slopes and embankments*

design situation for which the observational method is particularly appropriate. Settlements and pore pressures, and maybe lateral deformation as well, should be measured and recorded to monitor the performance of the structure during construction when using this method. Details about the use of the observational method are given in §2.8.

The surfaces of slopes may be protected against erosion either by sealing, using a geomembrane or some form of solid surface protection, or by planting, using grass, shrubs or trees. It is recommended in **C9.4(4)** that trees and shrubs should not generally be planted on river and lake embankments. This is to avoid the roots of trees entering rivers or lakes causing obstructions and because trees can lean towards rivers or lakes causing instability of embankments.

9.4 Ultimate Limit State Design

9.4.1 Slope Stability Analyses

As shown in Table 9.1, the first limit state that needs to be checked in the design of slopes is the ultimate limit state of loss of overall stability. It is envisaged in EC7 that slope stability analyses will normally be carried out using limiting equilibrium calculations, assuming that the failing slope consists of a single rigid block or several rigid blocks bounded by failure surfaces. However stability analyses based on statically admissible stress fields or finite elements are also allowed [**C9.5.1(2)**].

Where a harder stratum exists close to the surface or where the soil is granular, with no cohesion, the failure mechanism may consist of a shallow translational slide, often referred to as an infinite slope slide. In situations where the length of the slide is limited, a wedge analysis involving a series of wedges sliding over each other may be appropriate. For most ground conditions, particularly where the ground is homogeneous or where it has some cohesion, it is normally sufficient to assume that failure occurs through a rotational slide along a circular slip surface. For layered soils with weaker layers, it may be necessary to analyse the stability of the slopes failing along non-circular slip surfaces passing through the weaker layers.

EC7 requires in **C9.5.1(3)P** that, for any possible failure surface, equilibrium is checked when the actions and the soil shear strength parameters have their design values. As unsupported and unreinforced slopes are situations where only the strength of the ground is involved and there is no involvement of any structural material, Case C is the relevant design case. Thus the design self weight of the ground and the design groundwater pressures in slope stability analyses are equal to the characteristic values, since $\gamma_G = 1.0$ for case C. If there are variable loads, for example on the surface at the top of the slope, these are multiplied by $\gamma_q = 1.3$. The design soil strength parameters are obtained by dividing the characteristic values by the appropriate Case C γ_m values given in Table 2.6. If stabilisation measures incorporating structural elements such as retaining walls, anchors, nails or reinforced earth, are to be included, then Case B also needs to be considered.

No guidance is provided in EC7 on the suitability of drained and undrained analyses for assessing slope stability. Normally the long term performance of slopes is the critical design situation and hence drained analyses are appropriate, but undrained conditions are critical in the case of embankments on soft ground. Drained analyses should be carried out for the ULS groundwater conditions (see §2.5.2 and §3.2.5). As deformations may develop with time in cohesive soil slopes, the effective cohesion is likely to be small and unreliable and hence cautious values should be used for c'_k. Normally undrained analyses should only be considered for the design slopes in cohesive soils where short term stability is required, as in the case of excavations for foundations and trenches for services. However the length of time for which cohesive soils will behave as undrained will depend on the nature of the soil and its stress history and this needs to be assessed as part of the design process to determine if undrained analyses are appropriate.

9.4.2 Translational Slides

In a translational slide, failure occurs along a slip surface parallel to the ground surface, as shown in Figure 9.1, where the inclination of the slope is β, the depth of the slip surface is z and the unit weight of the soil is γ. Stability of the slope is analysed by considering equilibrium of a slice of the sliding soil of width $b\cos\beta$ as shown in Figure 9.1. The forces acting on this slice are the design disturbing force S_d, which is the component of the design weight of the slice acting parallel to the slip surface, and the design resistance on the slip surface, R_d. The equilibrium requirement is that, at the ultimate limit state, the design disturbing force does not exceed the design resistance on the slip plane; i.e. $S_d \leq R_d$. It is assumed that the interslice forces on either side of the slice are equal and opposite and so cancel out.

Disturbing force
For no variable loads on the surface, the design disturbing force is given by:

$$S_d = \gamma z \cos\beta \sin\beta = \gamma z b \frac{\sin 2\beta}{2} \tag{9.1}$$

Undrained conditions
For undrained conditions, the design resistance is:

$$R_d = \left(\frac{c_{uk}}{1.4}\right) b \tag{9.2}$$

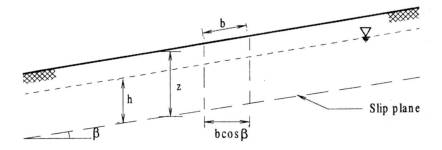

Figure 9.1: *Translational Slide*

Combining Equations 9.1 and 9.2, the critical slope for undrained conditions is obtained from:

$$\sin 2\beta = \frac{2c_{uk}}{1.4\gamma z} \qquad (9.3)$$

Drained conditions
For drained conditions, with the groundwater level parallel to the ground surface at a height h above the slip surface and steady state seepage parallel to the ground surface, the design resistance is:

$$R_d = (\gamma z - \gamma_w h) b \cos^2\beta \, \frac{\tan\phi'_k}{1.25} \qquad (9.4)$$

Combining Equations 9.1 and 9.4 the critical slope for drained conditions is obtained from:

$$\cos\beta \sin\beta = \left(\frac{\gamma z - \gamma_w h}{\gamma z}\right)\cos^2\beta \, \frac{\tan\phi'_k}{1.25} \qquad (9.5)$$

or

$$\tan\beta = \left(1 - r_u \sec^2\beta\right)\frac{\tan\phi'_k}{1.25} \qquad (9.6)$$

where the pore pressure ratio, r_u equal to the pore pressure on the slip plane divided by the total stress and is given by:

$$r_u = \frac{\gamma_w h \cos^2\beta}{\gamma z} \qquad (9.7)$$

When the groundwater level is at the ground surface, so that h = z in Equation 9.5, the critical slope is found from:

$$\tan\beta = \left(\frac{\gamma - \gamma_w}{\gamma}\right)\frac{\tan\phi'_k}{1.25} \qquad (9.8)$$

As $(\gamma - \gamma_w)/\gamma \approx 1/2$, the critical slope angle for this case, $\beta \approx \phi'_k/2.5$.

9.4.3 Rotational Slides

For rotational slides on a circular failure surface, the equilibrium condition is that the design disturbing moment shall not exceed the design resisting moment. For undrained conditions this gives:

$$W_d x \le \frac{c_{uk} L r}{1.4} \tag{9.9}$$

where:

W_d = design weight of the sliding mass and any surcharge loads;
x = distance of the centroid of the sliding mass from the centre of rotation;
L = length of the slip surface;
r = radius of the slip surface.

In a rotational slide, where the soil mass is sliding along a circular slip surface, most of the weight of the sliding soil, W_u acts unfavourably, causing sliding, while some of the weight, W_f acts favourably, resisting sliding as shown in Figure 9.2. In practice it is not feasible to distinguish between the favourable and unfavourable components of the soil weight. Instead the whole mass of the soil is treated as being from a single source and hence, as the overall effect is unfavourable, both parts are multiplied by the same partial factor for unfavourable permanent loads which, for Case C, is $\gamma_G = 1.0$.

9.4.4 Method of Slices

The method of slices may be used to analyse the stability of slopes in soils that do not exhibit marked anisotropy [**C9.5.1(4)**]. Although **C9.5.1(4)** is not presented as a Principle, it requires that, when using the method of slices, both the vertical and moment equilibrium of the sliding mass shall be checked and, if horizontal equilibrium is not checked, then the interslice forces shall be assumed to be horizontal. While no specific calculation methods are mentioned, these requirements mean that Morgenstern and Price's (1965) general analysis, in which all the equilibrium conditions are satisfied, and Bishop's (1955) methods, in which the resultant forces on the sides of the slices are assumed to be either horizontal or inclined, are acceptable. Janbu's (1957) method with inclined interslice forces is acceptable, however these requirements rule out the simpler methods, such as Janbu's (1957) method with horizontal interslice forces and Fellenius' (1927) method in which the interslice forces are neglected and vertical and horizontal equilibrium are not both satisfied.

9.4.5 Factors of Safety

The conventional practice in assessing the stability of embankments and slopes is to calculate an overall factor of safety, FOS in terms of the soil shear strength.

Translational slides

In the case of translational slides, the FOS equals the ratio of the characteristic resisting force to the component of the weight of the sliding mass acting parallel to the slip plane. For undrained conditions this is:

$$FOS = \frac{2c_{uk}}{\gamma z \sin 2\beta} \tag{9.10}$$

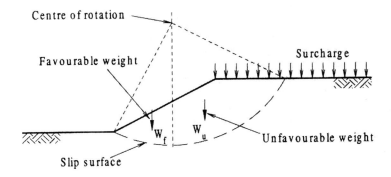

Figure 9.2: *Rotational slide*

which is equal to the partial factor on c_{uk}. Therefore, using the partial factor of 1.4 on c_{uk} is equivalent to adopting an overall factor of safety of 1.4.

For drained conditions with c' = 0, the FOS may be expressed as:

$$FOS = \left(1 - r_u \sec^2 \beta\right)\frac{\tan \phi'_k}{\tan\beta} \tag{9.11}$$

When compared with Equation 9.6, it is evident that the FOS is equal to the partial factor on $\tan\phi'_k$, i.e. 1.25, since the partial factor on the unit weight of both the soil and the water, and hence on r_u, is 1.0.

Rotational slides

For a rotational slide, the overall factor of safety in an undrained stability analysis is equal to the ratio of the characteristic restoring moment to the disturbing moment. From Equation 9.9 it is evident that, as for the case of a translational slide, the FOS is equal to 1.4, the value of the partial factor on c_{uk}, when there is no surcharge load on the surface.

The overall factor of safety in the case of the rotational failure of a drained slope in soil with cohesion as well as friction and a surcharge load (variable action) will normally exceed 1.3. This value is similar to the value commonly adopted in traditional slope stability designs using unfactored parameters values; for example BS 6031 (1981) recommends an FOS value between 1.3 and 1.4. However, without cohesion and no surcharge load, the FOS will be less than 1.3 and so less than the value traditionally used in the design of slopes.

To investigate the FOS of a slope against rotational failure, a 10m high drained slope in soil with $\gamma_k = 18kN/m^3$, $c'_k = 10kPa$, $\phi'_k = 30°$ and the groundwater level 1m below the surface, was analysed for Case C conditions. For no surcharge load at the top of the slope, the design slope angle was found to be 21.4°, giving an FOS of 1.37 whereas, with a surcharge load of 15kPa at the top of the slope, the design slope angle was found to be 27.2° giving an FOS of 1.39. The parameter values used in the design of a similar slope in Example 9.3 are shown in Table 9.4.

9.4.6 Excessive Ground Movements

In the ULS design of slopes, as shown by the list of limit states in Table 9.1, it is necessary not only to check against loss of overall stability but also to consider the possibility of the failure of supported structures or services due to excessive ground movements. Ground movements leading to an ULS are more likely to occur in the case of embankments on soft ground. In the case of an embankment it is necessary to check the movements within the embankment as well as in the ground beneath.

According to **C9.5.2(3)**, the settlement of embankments on soft ground may be calculated using the principles in **C6.6.1** for calculating the settlements of foundations. This means that the settlement of the soft ground should take account of the three components: undrained (elastic) settlements, consolidation settlements and creep settlements.

As noted in **C9.5.2(3)**, the calculation methods available at present do not usually provide reliable predictions of the pre-failure deformations of a slope. The occurrence of an ultimate limit state due to excessive ground deformations may therefore be avoided by either:

- limiting the mobilised shear strength by using higher partial safety factors; or
- observing the movements and taking action to control them if this becomes necessary. This is an example of the use of the observational method described in §2.8.

No guidance is given in EC7 on how much the mobilized shear strength should be limited in order to avoid excessive deformations of slopes. Some guidance, however, may be obtained from the overall factors of safety mentioned above that have traditionally been used for the design of slopes and embankments.

9.5 Serviceability Limit State Design

C9.6(1)P requires the design to show that, under the design actions, the deformation of a slope or embankment will not cause loss of serviceability in structures, roads or services sited on or near the embankment or slope. The text in **C9.6** on deformations causing serviceability limit states is similar to that in **C9.5.2** for excessive deformations causing ultimate limit states. As in the case of deformations causing ultimate limit states, reliable calculation methods are not usually available at present to predict the deformations of slopes causing serviceability limit states. According to **C9.6(2)**, the methods given in **C9.5.2(3)** for calculating the settlement of an embankment causing an ultimate limit state are also applicable for checking that a serviceability limit will not occur. According to **C9.6(2)**, the use of trial embankments may be useful for predicting the behaviour of embankments. As noted in Table 9.1, deformations due to changes in the groundwater conditions in an embankment or in the ground beneath an embankment should also be considered.

9.6 Slope Design Examples

The following slope design examples include stability analyses for translational and rotational slides for undrained and drained conditions. The slopes failing by rotational mechanisms have been analysed with the OASYS program, SLOPE using Bishop's method of slices with variably inclined interslice forces.

Example 9.1: Translational Slide, Undrained Conditions

The first example is the analysis of the stability of the long slope at an inclination of 30° shown in Figure 9.3. The ground has a unit weight of 18.7kN/m³ and there is a weak clay layer with c_{uk} = 25kPa at 1.8m depth and parallel to the ground surface.

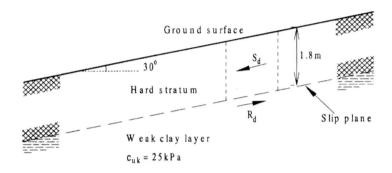

Figure 9.3: *Translational slide on a weak clay layer, undrained conditions*

Design situation and analysis
The stability of this slope is checked against failure by a translational slide along the weak clay layer for undrained conditions for Case C. Substituting the given parameter values into Equation 9.1, the design disturbing force, S_d is:

$$S_d = \gamma z b \frac{\sin 2\beta}{2} = 18.7*1.8*b*\frac{\sin(2*30)}{2} = 14.6b \text{ kN/m}$$

while, from Equation 9.2, the design resistance, R_d is:

$$R_d = \left(\frac{c_{uk}}{1.4}\right)b = \left(\frac{25}{1.4}\right)b = 17.9b \text{ kN/m}$$

Conclusion
As $S_d \leq R_d$ (14.6b < 17.9b), this slope has adequate resistance against a translational slide failure.

Example 9.2: Rotational Slide, Undrained Conditions

The stability of the 10m high clay slope, shown in Figure 9.4, inclined at 1 vertical to 2 horizontal, which equivalent to 26.6°, is to be analysed. The soil properties are γ = 18.4kN/m³ and c_{uk} = 42kPa and there is a uniformly distributed load (UDL), i.e. a variable load, of 20kPa on the surface at the top of the slope.

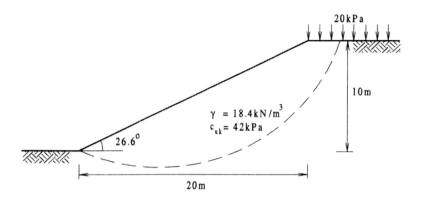

Figure 9.4: *Rotational slide, undrained conditions*

Design situation and analysis
The stability of the cutting is analysed for undrained conditions just for Case C for the most critical circular slip surface passing through the toe of the slope. The design parameter values are:

c_{uC} =   42 / 1.4   =   30.0kPa
UDL_d =   1.3 * 20   =   26.0kPa

Analysing the stability of the slope for these parameter values using the SLOPE program, it is found that $R_d = S_d$ and so Equation 9.9 is satisfied. The overall factor of safety, based on an analysis using unfactored parameters, is 1.44, which represents the combination of the partial factors on c_{uk} and the UDL.

Example 9.3: Rotational Slide, Drained Conditions
The stability of the 10m high slope, shown in Figure 9.5, inclined at 1 vertical to 2.5 horizontal, which equivalent to 21.8°, a UDL of 10kPa on the top surface is to be analysed. The soil properties are $\gamma = 19.0kN/m^3$, $c'_k = 15kPa$ and $\phi'_k = 28°$. the water table is 1m below the ground surface.

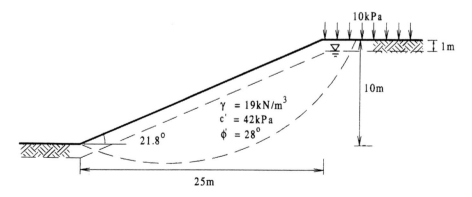

Figure 9.5: *Rotational slope failure, drained conditions*

Input parameters	Case B	Case C	Case C, $c'_k = 0$
γ_d	25.7kPa	19kPa	19kPa
γ_{wd}	13.5kPa	10kPa	10kPa
UDL_d	15kPa	13kPa	13kPa
c'_k	10kPa	10kPa	0
c'_d	10kPa	6.25kPa	0
ϕ'_k	28.0°	28.0°	28.0°
ϕ'_d	28.0°	23.0°	23.0°
Actual slope	21.8°	21.8°	21.8°
Results obtained			
Calculated R_d/S_d	1.37	1.03	0.72
FOS = R_k/S_k	1.39	1.39	0.91
Critical slope	28.6°	22.2°	15.5°

Table 9.4 *Parameters used in slope stability analysis in Example 9.3 and results obtained*

Design situation and analysis
The stability of the slope is analysed for drained conditions assuming a circular slip surface passing through the toe. The data and results of the SLOPE analyses for Cases B and C, and also for Case C when $c'_k = 0$, are presented in Table 9.4. As R_d/S_d for Case C is 1.03 and for case B is 1.37, these results clearly demonstrate that Case C controls the design and hence Case B is not relevant in a slope stability analysis not involving structural strength. The overall factor of safety, FOS is equal to 1.39 due to the combination of the Case C factors on c'_k, $\tan\phi'_k$ and UDL_k. This value would normally be considered acceptable. However, it may not be prudent to rely on c'_k. If the soil loses its effective cohesion, so that $c'_k = 0$, then the value of R_d/S_d for the actual slope of 21.8° reduces to 0.72, the FOS becomes 0.91 and the safe design slope reduces from 22.2° to 15.6°. This demonstrates the significance of variation in c'_k in slope stability analyses.

References

Bishop A W (1955) The use of the slip circle in the stability analysis of slopes, *Géotechnique*, 4:2.

BS 6031 (1981) *Code of practice for earthworks*, British Standards Institution, London.

Fellenius W 1927 *Erdstatische Berechnungen*, (Revised ed. 1936) W. Ernst & Sohn, Berlin.

Janbu N. (1957) Earth pressure and bearing capacity calculations by generalized procedure of slices. *Proceedings IV International Conference on Soil Mechanics and Foundation Engineering*, 2, 207-212.

Morgenstern N R and Price V E (1965) A Numerical method for solving the equations of stability of general slip surfaces. *Computer Journal*, 9:388-92.

Symbols and Abbreviations

Symbols

A	Spread foundation base area or pile shaft or pile base area	R	Resistance
A'	Effective foundation area	R	Bearing resistance
B	Foundation width	S	Sliding resistance
B'	Effective width of a foundation	U	Water pressure force
C	Serviceability criterion (e.g. limiting movement)	V	Vertical load or force
		W	Weight
C_c	Compression index	X	A soil property
C_s	Swelling index		
D	Downdrag force	a	Geometric data
D	Thickness of a spread foundation	b	Base inclination factor
E	Action effect	c'	Cohesion intercept in terms of effective stress
E	Young's modulus	c_u	Undrained shear strength
E_m	Drained Young's modulus in adjusted elasticity method	c_v	Coefficient of consolidation
		d	Depth
E_m	Young's modulus from cone penetration or weight sounding tests	d	Sheet pile penetration depth
		d	Spread foundation depth factor
E_{oed}	Oedometric modulus (one-dimensional compression)	e	Eccentricity
		f	Coefficient in adjusted elasticity equation for foundation settlement
E_M	Pressuremeter modulus	h	Height
F	Axial or transverse load or force on a pile	i	Spread foundation load inclination factor
F	Ratio of characteristic spread foundation resistance to characteristic loading	k	Permeability
		k_s	Coefficient of subgrade reaction
G	Permanent action or load	m_v	Coefficient of volume compressibility
G_s	Specific gravity of soil particles	n	Porosity
H	Horizontal action or force	p	Serviceability limit state bearing pressure on foundation base in adjusted elasticity method
I_D	Density index		
K	Coefficient of earth pressure		
L	Length of a spread foundation	p_{LM}	Pressuremeter limit pressure
L'	Effective length of a spread foundation	q	Total overburden or surcharge pressure at level of foundation base
M	Moment, bending moment		
N	Bearing resistance factor	q	Pile resistance per unit area
N	SPT blowcount	q'	Effective overburden or surcharge pressure at level of foundation base
N_{10}, N_{20}	Dynamic probing test blowcounts for 10 and 20mm penetration		
		q_c	Unconfined compression strength
Q	Variable action or load	r_d	Dynamic probe test

s	Settlement	γ	Soil unit weight
s	Spread foundation shape factor	γ_F	Partial load factor
t	Thickness	γ_m	Partial material factor
w	Water content	γ_R	Partial resistance factor
w_L	Liquid limit	ρ	Density
w_P	Plastic limit	σ'	Effective normal stress
		ξ	Conversion/correlation factor (xi)
Δ_a	Geometric safety margin	ϕ'	Angle of shearing resistance in terms of effective stress
α	Angle of inclination of spread foundation base		
δ	Angle of shearing resistance between ground and structure	ψ	Coefficient for combination value of a variable action (psi)

Subscripts

A	Case A	oed	Oedometer (e.g. E_{oed})
B	Case B	p	Passive earth pressure
C	Case C	q	Bearing resistance component due to surcharge
DMT	Dilatometer test		
PLT	Plate loading test	s	Serviceability limit state (e.g. C_s)
a	Active earth pressure	s	Pile shaft
a	Anchorage	t	Tension
a	Pile shaft adhesion	t	Total (e.g. total pile resistance)
b	Pile base	tr	Transverse
c	Bearing resistance component due to cohesion	u	Undrained
		u	Ultimate limit state (e.g. R_u)
c	Compression pile	v	Vertical
d	Design		
dl	Rock dilatometer test	0	At rest or initial condition
h	Horizontal		
k	Characteristic	γ	Bearing resistance due to soil effective weight
m	Measured (e.g. pile resistance)		

Abbreviations

CBR	California bearing ratio	GIR	Geotechnical investigation report
CPT	Cone penetration test		
CPTU	Piezocone test	MCV	Moisture condition value
DMT	Dilatometer test	PLT	Plate loading test
FOS	Overall factor of safety	PMT	Pressuremeter test
GC1	Geotechnical Category 1	SLS	Serviceability limit state
GC2	Geotechnical Category 2	SPT	Standard penetration test
GC3	Geotechnical Category 3	ULS	Ultimate limit state
GDR	Geotechnical design report	WST	Weight sounding test

Index

00965 8923

9 781447 112068